D0312789

CLIMATEGATE

CLIMATEGATE

A VETERAN METEOROLOGIST
EXPOSES THE GLOBAL WARMING SCAM

BRIAN SUSSMAN

WND Books

CLIMATEGATE
WND Books

Published by WorldNetDaily
Washington, D.C.

WRITTEN BY BRIAN SUSSMAN
JACKET DESIGN BY MARK KARIS
INTERIOR DESIGN BY ELYSE STRONGIN, NEUWIRTH & ASSOCIATES, INC.

WND Books are distributed to the trade by:
Midpoint Trade Books
27 West 20th Street, Suite 1102
New York, NY 10011

WND Books are available at special discounts for bulk purchases. WND Books, Inc.
also publishes books in electronic formats. For more information call
(541-474-1776) or visit www.wndbooks.com.

First Edition

ISBN: 978-1-935071-83-9

Library of Congress information available

Printed in the United States of America

10 9 8 7 6 5 4 3 2

ACKNOWLEDGEMENTS

I COULD NOT have spent two years crafting this book without the faithful encouragement of the woman I have loved since I was a teenager, my beautiful, sensible, and inspirational wife. For many years she encouraged me to write this book. Finally, with her embracing support, as well as the backing of our outstanding children, I committed myself to devote hundreds of hours to get it done. Thank you, dear family, for cutting me loose to make this happen.

I am also grateful for my wonderful agent, the wise and gracious Rachelle Gardner at WordServe Literary Agency. Rachelle, with little more than a synopsis and a couple of completed chapters, you believed in this project. I am blessed by your representation.

Likewise, I am honored to have Joseph Farrah and his excellent team at WND Books as my publisher. Because of your truth-telling, Joseph, this book will embolden an audience hungry for the truth.

Hearty acknowledgements must be given to El Rushbo, for motivating me to trade the restrictive environment of the television newsroom for a liberating career in talk radio; Mark Levin, for emboldening my worldview; Mickey Luckoff and Jack Swanson, for equipping me to host a successful program whereby I'm able to share my research and opinions daily; Lee Troxler, for many hours of writing mentorship; and to my dear friend John Frattarola, for his faithful service as my editor, advisor, and "literary pastor." John, you made this book much better. Thank you.

To the many people who pushed me forward with their prayers and uplifting words of cheer (including my beloved "growth group"), I am indebted to you. To those who discouraged me from writing this book, you drove me to work even harder. Young Aaron, thanks for digging up a few of the gems found within these pages. Brother Brad, thanks for checking my work. Golf Gazoo, thank you for laying down your clubs long enough to confirm my calculations. And "Dave," I am especially beholden for you reaching out to reveal Al Gore's "web of fortune." Your identity remains my secret.

In addition, I must recognize the desperate climate scientists/activists whose damning, cover-up email messages were discovered as this book was in final production. Your futile efforts to conceal your fraudulent claims that the earth is about to warm past the point of no return serves to verify my conclusion that you have bastardized the scientific method and have been fronting the biggest scam in history. Your ridiculous internal messages sure added some spice to this book.

Finally, Mom, thanks for being my biggest fan.

And to the memory of my dad, a man who—as you will discover—never put up with any bull.

CONTENTS

FOREWORD

... for us man himself is mutually of no value.

—Karl Marx, 1844

GLOBAL WARMING'S STORY begins with a diabolical bastard named Karl Marx.

Born in Germany in 1818, Marx lived sixty-five years during which time his twisted mind conceived an atrocious plot to infect the world with his godless philosophy of "organized collectivism"—a.k.a. communism, or, for the more politically correct, socialism.

Since his death in 1883, the global carnage wrought in his name by committed devotees is unfathomable. Commencing with the Russian Revolution in 1917 to the present, Marx's blood-red ideology has been responsible for the documented deaths of over 110 million individuals around the world. Hundreds of millions more have been forced to live in oppressed societies, void of the unalienable rights of life, liberty, and the pursuit of happiness.

Finally, after decades of stealthy determination, the quixotic conjectures of Marx have seeped into the framework of the United States, with the most effectual being the supposed environmental crisis known as global warming or climate change.

My name is Brian Sussman. I am a devoted husband, dedicated dad, conservative radio talk host in ultraliberal San Francisco, and a former award winning television meteorologist and science reporter. I have written this book to sound a vociferous warning: global warming is a scam perpetrated by an elite sect of Marx-lovers who believe they can do communism/socialism more effectively than their predecessors; and now, with the ascendance of Barack Obama as president, the scam has reached hyperspeed.

The information carefully detailed in this book will allow you to understand the political and monetary motivations behind this grand scheme, and will enable you to thoroughly wrap your brain around the science that is being purposefully abused. By the end of *Climategate* you will be able to explain to those who have innocently bought into the global warming lie that they've been swindled. Likewise, this treatise will provide you with the confidence to confront those who are driving the notion of climate change as a vehicle to bring forth the machinations of Marx.

MARXISM'S CURRENT APPEAL

It is critical to note that Karl Marx and his co-conspirator and writing partner, Friederich Engels, tickle modern day, pseudo-intellectual, elitist ears, in large part because their fiendish philosophy is intriguingly metaphysical.

To Marx and Engels, matter—atoms, molecules, and the otherwise unseen—is all there is or ever will be. Matter is the alpha and omega of their reality, providing the complete explanation for plants, animals, man, consciousness, intelligence, planets and solar systems. Marx and Engels held that "if science can get to know all there is to know about matter, we will then know all there is to know about everything."[1]

In other words, the founders of collectivism trusted in an amoral system of knowledge, as opposed to a moral structure founded on belief in God. Such a patrician worldview resonates with today's elites. Through advanced education and absorption of information, they surmise they can become masters of the universe. There is no room for absolute truth in their camp, only relativism, lest they be forced to acknowledge a divine being. To them, God is a mythological crutch for imbeciles, and whatever

---

1 Friedrich Engels, *Ludwig Feuerbach* (1886).

morality is required shall be defined by them, subject to their goals. Thus, telling a lie or perpetrating a fraud is perfectly acceptable—as long as the end justifies the means.

Confirming this secular system are Marx and Engels' Three Laws of Matter. These "laws" are to the Marxist, what the Ten Commandments are to a believer in God. They are the Law of Opposites, the Law of Negation, and the Law of Transformation. Together, these planks provide the rationale for the radical environmental policies being instituted today.

The Law of Opposites supposedly illustrates how everything in existence is a combination of diametrics, working in unity. For example, electricity is characterized by a positive and negative charge. Atoms include protons and electrons, which are contradictory forces working in unity. Even the human race is composed of opposite qualities: altruism and selfishness, courage and cowardice, humility and pride, masculinity and femininity. To function properly, Marx believed, these opposite forces must be kept in balance; if not, discord is certain. Thus, the Law of Opposites demands that humans must be kept in check because, as the most advanced creature, they can wreak the most havoc—hence, the need for a tightly regulated, often heavy handed, system of government. This is why American elites slobber over Castro's Cuba and herald Venezuela's president for life, Hugo Chavez, as heroes. According to the Law of Opposites, demagogues are essential to effective, domineering governance.

The Law of Negation adds a Malthusian component to Marx's madness. It declares that all nature is constantly expanding through death. To support this law, Engels created an awkward illustration using barley seed, which germinates—via its own death or negation—and produces many new plants. This clumsy concept seems especially cockamamie when applied to the human race, but Engels rationalized it by claiming, "... out of this dynamic process of dying the energy is released to expand and produce many more entities of the same kind."[2]

Marxists believe that Negation implies that all living matter has an inherent tendency to grow out of control—particularly humans. Thus, systems must be put into place to maintain sustainability, including mechanisms to assure human population control when necessary.

---

2 Friedrich Engels, *Anti-Duhring* (1878).

The third axiom is the most arrogant of this triad: the Law of Transformation. Transformation states that a continuous quantitive development by a particular species often results in a "leap" in nature, whereby a completely new form or entity is produced. This law was bolstered by the findings of Marx and Engels' contemporary, Charles Darwin. Darwin's theory of evolution sealed the communist founders' convictions that such "leaps" not only allowed for the origin of new species, but a leap *within* a species—particularly *homo sapiens*—could enable some to advance to new levels of reality. In other words, occasionally some people are spit out of the womb with a superior brain, making them more worthy than others. Thus, the Law of Transformation confirmed an elite status within the human race; and those born into evolution's aristocracy possess a duty to dictate how the underdeveloped shall live.

With these new revelations, Marx and Engels arrogantly boasted that "the last vestige of a Creator external to the world is obliterated."[3]

MARX'S WORST NIGHTMARE: AMERICA

Karl Marx began his rebellion against God as a student at the University of Berlin, where he was strongly influenced by the philosophy of Georg Hegel and a group of his followers known as the Young Hegelians. Their goal was straightforward: liquidate Christianity. In 1841, Marx received a doctorate in philosophy.

Karl Marx and his band of rebels were well aware of the republican form of government being tried in America—and loathed it. Marx perceived America's founders as reckless, Christian simpletons who were peddling dangerous propaganda—especially when they placed the phrase, *life, liberty and the pursuit of happiness* into their Declaration of Independence. To the collectivist, such freedoms were—and are—preposterous. The *life* of an individual is not unique—just a fragment of the ever-multiplying collective mass—the result of a random, cosmic, Darwinian accident. Likewise, *liberty* is an unattainable notion. The human masses are wholly incapable of coexistence without formidable government control and

---

3 Ibid.

regulation. Furthermore, the *pursuit of happiness* is the most egregious maxim of all, and Marx was aware of the origins of this key phrase. It was penned in direct reference to the words of English philosopher John Locke, who in his 1690 essay, *Concerning Human Understanding* wrote, "the necessity of pursuing happiness [is] the foundation of liberty."

Locke further explained:

> God, when he gave the world in common to all mankind, commanded man also to labour, and . . . subdue the earth, i.e. improve it for the benefit of life, and therein lay out something upon it that was his own, his labour. He that in obedience to this command of God . . . thereby annexed to it something that was his property, which another had no title to, nor could without injury take from him.
>
> —*Second Treaty on Civil Government*, 1690

America's founders understood that property was synonymous with liberty and security. They comprehended that in a capitalistic, free-market economic system void of overreaching government regulation, new workers or immigrants could progress up the class ladder in conjunction with their effort, owning their own business, farm, home, and estate. Such beliefs were offensive to Marx. The concept of a God that would command humans to take ownership of land, improve it, and even defend that property as their very own was hopelessly flawed and flew in the face of his Laws of Matter.

As an antidote to the presumptuous experiment being conducted in the United States, in 1849 Marx and Engels presented to the world their final formula for revolution, which they called the *Manifesto to the World*. The infernal document would eventually be known as *The Communist Manifesto*.

In chapter two of his Manifesto, Marx boldly states the goal of his envisioned new world order: ". . . the theory of the Communists may be summed up in the single sentence: Abolition of private property."

The perverted musings of these original community organizers opened both the bloody floodgates of tyranny and the green gates of environmentalism.

COMMUNISM'S BLOODY HANDS

Engels died in 1895, twelve years after Marx's death. In autumn of '95, a double obituary was penned by a young, dedicated disciple of both, Vladimir Lenin. In it he wrote:

> In their scientific works, Marx and Engel were the first to explain that socialism is not the invention of dreamers, but the final aim and necessary result of the development of the productive forces in modern society. All recorded history hitherto has been a history of class struggle, of the succession of the rule and victory of certain social classes over others. And this will continue until the foundations of class struggle and of class domination—private property and anarchic social production—disappear. The interests of the proletariat demand the destruction of these foundations, and therefore the conscious class struggle of the organized workers must be directed against them. And every class struggle is a political struggle.[4]

Inspired by his pedagogues, Lenin grasped the communist baton, ready to level the playing field and vaporize the ability of those who desired to move up the economic ladder, own property, and live economically self-sufficient. In his zeal to carry out this mission, he articulated a dearly held concept of the end justifying the means: "We say that our morality is wholly subordinated to the interests of the class struggle of the proletariat."[5]

To destroy the foundations of property and a capitalistic economy, every option had to be on the table: lying, cheating—even murder—were necessary tools.

In 1917 Lenin took power after the earlier overthrow of Russia's Czarist regime, and the official reign of communist terror began with a simple red flag flown heralding victory. That blood red flag became the color of communism, lasting to this day. Lenin's Russia would transition to Stalin's Soviet Union. Much of Europe, Asia, and Africa would eventually collapse to the communist/socialist ideology. During the ensuing 70 years,

---

4 Vladimir Ilyich Lenin, Obituary marking the death of Frederick Engels, August 1895.
5 Vladimir Ilyich Lenin, *Collective Works*, vol. 31 (Moscow, 1966), 291.

over 110 million people would be exterminated around the world in the name of Marx'ss better way . . . after all, the Laws of Matter and the Manifesto don't prohibit the spilling of blood—*they excuse it.*

MARX IN AMERICA

The tentacles of Marxism have been steadily reaching into the United States for decades. Red-flag-waving propeller-heads in academia faithfully undermine American values in the classroom; useful idiots associated with a variety of nongovernmental organizations work as tireless, agitating activists; and slimy, liberal politicians craftily seek ways to undermine constitutional principles through, as stated by Lenin, political means. Working in lockstep with the unseen communist bureaucrats, ensconced in the United Nations complex overlooking the East River in New York, this cabal of collectivists have discovered the ultimate tool to force social change upon first America, then the entire world: the very air we breathe.

Beginning with a series of fictitious ecology books in the Sixties, the institution of Earth Day in the Seventies, the rise of Al Gore in the Eighties, the Rio Earth Summit and subsequent U.N.-sponsored global warming scare campaign in the Nineties, and the relentless media onslaught proclaiming climate change in the new millennium, we have now been herded to the edge of a dangerous precipice; and, with the selection of Barack Obama as president, the powers that be are attempting to shove us over the brink.

Even before the election there were a multitude of clues regarding Obama's Marxist machinations. Certainly one appeared when, for the first time in history, a candidate's policies were referred to by his opponent as socialist. Barack Obama's only defense to Senator John McCain's charge that his "spread the wealth around" comment to Joe the Plumber was commensurate with socialism, was to clumsily reply, "by the end of the week he'll be calling me a secret communist because I shared my toys in kindergarten. I shared my, shared my [sic], peanut butter and jelly sandwich."[6]

Obama was wrong on both counts. Forcing folks to spread their wealth around *is* socialism, while sharing is akin to charity—not communism.

---

6 Barack Obama, campaign speech, Raleigh, North Carolina, October 29, 2008.

America should have taken a further cue when, just after the election, Sam Webb, Chairman of the Communist Party USA, stated in a speech that with Barack Obama in place, "an era of progressive change is within reach, no longer an idle dream."[7]

Finally, an unmistakable harbinger came with selection of President Obama's handpicked Director of the White House Office of Science and Technology Policy, John Holdren, who let the Marxian-striped cat out of the bag long ago when he wrote, "A massive campaign must be launched to restore a high-quality environment in North America and to de-develop the United States.... Resources and energy must be diverted from frivolous and wasteful uses in overdeveloped countries to filling the genuine needs of underdeveloped countries. This effort must be largely political."[8]

De-develop the United States to fill needs in *underdeveloped countries*, through *political* efforts. This is the conviction that emanates from the White House and is now racing through the veins of government, both federally and locally. An elite brigade of zealots has cleverly created a new political platform to carry out the collectivist goals of redistributing wealth and destroying personal liberty, utilizing something that Karl Marx himself never envisioned: the environment, or more specifically, *the climate*. And because the efforts are political, these egalitarians are willing to utilize phony science as terror tactics, in an attempt to force you to believe that your lifestyle is responsible for negatively altering the earth's atmosphere.

It's all a lie.

The earth is not warming, and climate always changes—and they know it.

But they are assuming there are enough ignorant fools they can hornswoggle into believing that any climatic alterations, including extreme weather events like hurricanes or floods, are being caused by mankind.

Global warming is the grandest of all tyrannical schemes.

---

7 "Off and Running: Opportunity of a Lifetime," speech by Sam Webb, Chairman, Communist Party USA, delivered at a conference hosted by *People's Weekly World*, Cleveland, Ohio, January 31, 2009. Transcript available at http://www.politicalaffairs.net/article/articleview/8085/.

8 Holdren, Ehrlich and Ehrlich, *Human Ecology: Problems and Solutions*, WH Freeman and Company, San Francisco, 1973, page 279.

1

FOUNDATION OF FRAUD

[W]e need to get some broad based support, to capture the public's imagination. That, of course, means getting loads of media coverage. So we have to offer up scary scenarios, make simplified, dramatic statements, and make little mention of any doubts we might have."

—Stanford Climatologist Stephen Schneider, author of *Global Warming*[9]

I SNIFFED THIS one out in 1970. My high school Earth Science teacher mandated our class "celebrate" the newly created Earth Day. I had always suspected this guy was a weirdo, and now those suspicions were confirmed. Just a few years out of teachers' college, his hair was parted down the middle and longer than the older male instructors. He sported the round, wire-rimmed glasses that had recently come into style with all the hippies, and of course, he wore a beard. To me, a jock, he was basically a flower-child with a teaching job, but a nice enough guy, nonetheless.

Smiling like it was Christmas, he walked about the room distributing green Earth Day buttons, mentioned something about this being a "teach-in" and urged us to pin the buttons on to "save the earth." We all did, of course, with hopes of the pin translating to a good grade in his class. Heck, with all the environmental disasters we'd seen on the nightly

---

9 Interview with Stephen Schneider, *Discover* magazine, October 1989.

news recently, who could be against clean water and air? But for me there was a problem. Deep down, for some reason I couldn't exactly put my finger on, the whole thing smelled like a gimmick, and I remember thinking, *How can wearing a stupid button save the earth?*

My dad, on the other hand, knew exactly how to put his finger on it.

"Bullshit!" he proclaimed when I came home wearing the button that night. His generation possessed quite a nose for a scam, and when they smelled it, had no reservations warning folks not to step in it.

"So, we're celebrating the earth now—kind of like a birthday," he said, looking at my Earth Day button, his words dripping with sarcasm.

I just shrugged, embarrassed.

"You're wearing that button and you don't even know what you're celebrating?"

"My science teacher gave them out."

"And you just pinned it on?"

Now I really felt like a goof. I didn't respond.

Without breaking a smile, my dad cracked, "You know, I'd love to purchase a gift for Mother Earth, but what do you buy for someone who has *everything?*"

It took a second, but when I saw that gleam in his eye, I cracked up. He joined in, slapped me on the shoulder, and with great manly pride, I removed the moronic button.

Looking back, I wish I had the insight to use my old man's line on my science teacher. However, for the many years I presented the weather on television in the otherwise ultraliberal San Francisco Bay Area, I used the line every April 22, and always got a hearty laugh from my co-anchors.

But honestly, Earth Day is no laughing matter.

WINDS OF CHANGE

I was unaware at the time, but high schools in the Seventies were radically changing, particularly public schools. They were incorporating the first assignments of what would eventually lead to the full-fledged social engineering centers that most have become today. Looking back on my four-year journey, the engineering was clearly the result of this new crop of flower-power teachers. Influenced by a myriad of seditious authors and revolutionary professors in college, these new educators spoke out freely against "the establishment" and "the man." Translated, that meant government, military, police, traditional religion, corporations, and parents—especially authoritative fathers—which certainly included dads like mine.

The book that seemed to resonate with these young teachers like no other was Rachel Carson's 1962 best-seller, *Silent Spring*, and, consequently, it became required reading for their students. When *Silent Spring* made it into my literature class, my antennae immediately rose. I can only attribute it to the skeptical spirit of my dad in me; by osmosis, I seemed to be absorbing his sense of smell.

Rachel Carson was a crafty wordsmith with a rudimentary training in zoology. While working as a writer for the U.S. Department of Fish and Game, she was influenced by many forerunners to the environmental movement and in crafting *Silent Spring* became queen of them all. The premise of her book was aimed at the United States: as a society, she believed, we were selfish and wasteful; our system of capitalism was inherently evil, and our businesses and corporations were knowingly raping the planet. Specifically, Carson dialed in on America's chemical manufacturers.

These greedy enterprises were in cahoots with the federal government, she claimed, and knowingly involved in creating and promoting the use of deadly chemicals—all for unseemly profit, of course. *Silent Spring*'s primary target was a compound in widespread use: dichloro-diphenyl-trichloroethane, a.k.a. DDT. DDT was a common compound used to rid neighborhoods, like mine, all across America, of pesky insects, especially disease-carrying mosquitoes. It was also widespread and effectively used throughout Africa and third world countries to eradicate those regions of deadly, mosquito-borne malaria. Carson's claim that the insecticide was lethal to man and animals, with the manufacturers' knowledge, was just too conspiratorial for me to believe. DDT, she said, devastated bird populations like the robin, peregrine falcon, and bald eagle; even the birds' eggs were thinning from the widespread use of the compound. And, once it was introduced into the food chain, Carson believed it would become a carcinogen to mankind.

If that were true, I wondered why these corporations, supposedly raking in millions in profits, would intentionally kill their customers and thus eliminate their consumer market? To me it made no sense.

However, Carson's clever writing, combined with the zeal emitted from the young, evangelistic teachers, was causing most of my classmates to swallow this stuff whole—but I wouldn't bite. Instead, I took just enough of a whiff of this bull to garner a decent grade on my subsequent book report. Earth Day and *Silent Spring* were my first tastes of environmental fraud.

I graduated high school, zipped through college, and quickly landed a job as a television weatherman. Eventually my performance on the air would be recognized with the esteemed and coveted American Meteorology Society Seal of Approval, a host of state and regional awards and even a couple of Emmys. It was a career I would enjoy for the next two decades.

LIES EXPOSED

In 1983, I was moving up the television news ladder and took a great weather gig in the San Francisco market. One evening, the station's anchorman told me about a captivating speaker he had just heard at a civic luncheon. The speaker was Dr. J. Gordon Edwards of San Jose State University, who began his talk by pouring a teaspoon of DDT into a glass of water and then drinking the glass empty. Edwards was a colorful entomologist who had taught at the university for decades. Besides being a bug doctor, he was also a famed mountaineer and well-known conservationist, who was making quite a name for himself by illustrating the lies of Carson's *Silent Spring*. It was everything I had been waiting over a dozen years to hear.

In 1962, Edwards was working with a team of researchers at Glacier National Park in Wyoming when he first read Carson's book, and he, too, smelled something noxious—especially regarding DDT.

Edwards knew that the compound dichloro-diphenyl-trichloroethane was first created by a German scientist in 1874, and that, in 1939, Swiss scientist Paul Muller, who eventually won a Nobel Prize for his work with DDT, perfected it as an effective, humanly-safe insecticide. In many ways, it was a miracle compound: inexpensive, nontoxic to humans, and extremely effective in exterminating targeted insects, while other forms of wildlife seemed to be immune to its effects. Edwards also had first-hand experience with DDT. In 1944, while serving as a U.S. combat medic stationed in Italy, his company was plagued with body lice. The fast-breeding insects were spreading typhus among the troops in some parts of Europe—frightening, given the fact that some 30 years prior typhoid had killed 3 million people in that very same part of the world. To curtail the developing epidemic, the chemists at Merck & Company in New Jersey produced the first 500 pounds of American-made DDT and rushed it to the troops in Italy, a move that would seem unlikely if DDT was a deadly nerve agent. Edwards's job was to dust the troops.

"For two weeks I dusted the insecticide on soldiers and civilians, breathing the fog of white dust for several hours each day," Edwards later wrote.[10]

The DDT worked as intended, and the typhus outbreak was checked. The Surgeon General estimated that the DDT saved the lives of at least 5,000 soldiers. Inspired by this experience, Edwards went on to get his Ph.D. in entomology from Ohio State University.

Now, years later, as an expert, Professor Edwards was methodically picking apart *Silent Spring* in his affable style. His extensive research completely debunked the primary pillars of Carson's book. For example, DDT is not a carcinogen[11] and in fact, might be a cancer-fighting drug.[12] (Despite his showy consumption of DDT, cancer never struck Dr. Edwards. He died in 2004 at the age of 84 while mountain climbing in his beloved Glacier National Park).

DDT does not thin bird eggshells. Experiments attempting to associate DDT with eggshell thinning involved mega-doses of the compound, far higher than would ever possibly be encountered in the wild.[13] Instead studies illustrate that thinned eggshells are caused by a number of natural factors including dehydration, old age, and extreme temperatures.[14]

---

10 J. Gordon Edwards, Mosquitoes, DDT, and Human Health, *21st Century* magazine, Fall 2002, http://www.21stcenturysciencetech.com/articles/Fall02/Mosquitoes.html.

11 K.C. Sillinskas & A.B. Okey, *Journal of the National Cancer Institute*, 55:653-657, 1975.

12 E.R. Laws, *Archive of Environmental Health*, 23:181-184, 1971.

13 Statement and affidavit by W.E. Hazeltine, EPA Hearings on Tussock Moth Control, Portland, Oregon, page 9, January 14, 1974.

14 R.K. Tucker and H.A. Haegle, *Bulletin of Environmental Contamination and Toxicology* 5(3): 191-194, 1970, (30 percent thinner shells were formed after quail were kept from

The death of wild birds due to feeding in marshes, agricultural fields, and forests treated with DDT is a total farce. During the years DDT was used in the United States there was a significant increase in the numbers of pheasants, quail, doves, turkeys, and other game species.[15] Despite their low numbers in the year in which *Silent Spring* was written, bald eagles were never threatened by DDT. Instead, because of reckless hunting, these iconic raptors were threatened with extinction as far back as 1921, 25 years before widespread use of DDT.[16] Comparing the Audubon Society's pre-DDT bald eagle count of 1941 to that of 1960, there was actually a reported 25% increase in the bird's population.[17] Likewise, the decline in the U.S. peregrine falcon population occurred long before the DDT years.

Unfortunately, Dr. Edwards's persuasive research hit the scene too late. In 1972, the pressure created by *Silent Spring* was the impetus for banning DDT domestically and internationally, eventually leading to the deaths of millions from malaria.

Tragically, today the late Rachel Carson is regaled as a hero. Forests, government buildings, and public schools have been named in her honor. Visit any book store, and you'll see a new copy of *Silent Spring* on the shelf, as it remains required reading for the drones who desire thorough environmental indoctrination. When confronted with the many lies contained within *Silent Spring*, Carson's committed disciples always stutter the same excuse: "but her *intentions* were good." But good intentions do not make for good science.

---

water for 36 hours); M.L. Sunde, *Farm Technology*, Fall 1971, (older birds produce thinner shells); Romanoff and Romanoff, 1949. *The Avian Egg*, Wiley & Sons, 1949 (temperature effects bird eggs).

15 C.F. Wurster, Department of Biological Sciences, State University of New York, Stony Brook, testifying before Congress, May 5, 1969.

16 W.G. Van Name *Ecology* 2:76, 1921.

17 Testimony of J.J. Hickey during Congressional hearings conducted before an EPA examiner regarding the effects of DDT, 1971; "The Myth of the Vanishing Peregrine Falcon: A Study in Manipulation of Public and Official Attitudes." Canadian Raptor Society Publication; "The 42nd Annual Christmas Bird Census," *Audubon Magazine* 44:1-75, January/February 1942; "The 61st Annual Christmas Bird Census," Audubon Field Notes 15(2):84-300, 1972.

MARX'S MEN AT STANFORD

A few years after Carson's book stole the minds of many, a second screed claimed more hostages in 1968—Paul Ehrlich's *Population Bomb*. Ehrlich, a professor at Stanford University, has authored many best-selling social engineering books during his decades at the liberal Bay Area institution, but the *Bomb* was his first hit. He has gained a reputation amongst environmentalists as a prophetic guru, and, even though his predictions have been both thoroughly outrageous and unattained, he remains an icon within their movement. "The battle to feed humanity is over," he proclaimed in *Population Bomb*. "In the 1970s and 1980s hundreds of millions of people will starve to death in spite of any crash programs embarked upon now."[18]

Playing straight from Marx's Laws of Matter, Ehrlich has long opined the earth is being forced to support too many people, who require too many resources and produce too much pollution. For Ehrlich, the solution has never been promoting honest governments and freer economies in the third world so that infrastructure could be constructed, clean water and sanitation readily available, and successful farming practices implemented. Ehrlich's conviction has always boiled down to people being the problem—there are just too damned many of them—and his final solution has always been clear: "Population control is the only answer."[19]

Ehrlich's book was also on the required reading list in my high school, and it's a good thing my old man didn't browse my copy of the *Bomb*, or he would have blown up. After all, not too many years earlier he helped fight a war to defeat a dictator who also believed in population control.

Ehrlich's wild allegations included equating the earth's surplus of people with a *cancer* that needs to be eradicated: "A cancer is an uncontrolled multiplication of cells; the population explosion is an uncontrolled multiplication of people. . . . We must shift our efforts from treatment of the symptoms to the cutting out of the cancer. The operation will demand many apparently brutal and heartless decisions."[20]

---

18 Paul Ehrlich, *The Population Bomb*, New York: Ballantine Books, 1968, Page 1.
19 Ibid.
20 Ibid., 166.

The method to Ehrlich's madness was revealed in his dictatorial action plan: "Our position requires that we take immediate action at home and promote effective action worldwide. We must have population control at home, hopefully through changes in our value system, but by compulsion if voluntary methods fail."[21]

Silent Spring convinced a generation that modern American liberty, ingenuity, free enterprise, and capitalism were the problem and ruining the planet. Piggybacking on Carson's success, Ehrlich introduced the solution: *a change in our value systems . . . by compulsion if other methods fail.*

Marx and Engels couldn't have executed a better one-two punch.

A new ethos was rapidly developing and the American ideals of Life, Liberty and the Pursuit of Happiness were under assault; in fact, on the final page of her book, Carson actually chides such notions as "Neanderthal."

ECOLOGY-INSPIRED DEATH

Observe the results achieved by the ghoulish guidance provided by Carson and Ehrlich. Since the DDT ban in 1973, malaria has killed 96 million people.[22] AIDS, a horrible disease that has taken a similar number, could be greatly reduced in the third world through abstinence education and the encouragement of monogamy, both of which are mocked by today's amoral, elitist policymakers. Famine, preventable through responsible government management and proper utilization of natural resources, kills millions more each year. Abortion, the termination method of choice

---

21 Ibid., xi–xii.

22 Based on an estimated 2.7 million malarial deaths per year—"The Intolerable Burden of Malaria: A New Look at the Numbers," supplement to *The American Journal of Tropical Medicine and Hygiene.* The supplement was published by the Multilateral Initiative on Malaria (MIM), including NIH, The Centers for Disease Control and Prevention, GlaxoSmithKline, the Rockefeller Foundation, the United Kingdom Medical Research Council, the United Nations Foundation, the United States Agency for International Development (USAID), the Wellcome Trust, and the World Health Organization. (Thanks to Steve Milloy, publisher, Junkscience.com, for his continually running "malaria clock").

for over 1 million pregnancies each year in the United States alone, is fully promoted and sanctioned by the government and medical profession alike. The math adds up to over 250 million people eliminated in less than 40 years.

And now, the eco-mentality first propagated by Carson and Ehrlich has become mainstream. A headline in Great Britain recently read, "Having Large Families is an Eco-crime."[23] The story told of women in England boasting of their abortions and sterilizations in the name of saving the earth.

"Having children is selfish," one fool told London's *Daily Mail.* "It's all about maintaining your genetic line at the expense of the planet."[24]

While such barrenness would likely be applauded by Ehrlich, he has recently come up with a new twist to sell his Malthusian snake oil: combining love of country with the number of children a couple may have.

"Patriotic Americans stop at two," he boasted to the *Christian Science Monitor* in 2009.[25]

It must also be noted, just as Marx had Engels, since the Seventies Ehrlich has relied on a trusted writing partner to co-author many of his books: President Obama's supreme science advisor, John Holdren. The duo became co-conspirators while teaching at Stanford, together penning some beauties, like pondering whether it was wiser to spend $1.8 billion to fund vasectomies or on a nuclear power plant that generates electricity without greenhouse gas emissions.[26]

The foundation of fraud laid in the Sixties and Seventies was a solid and lasting one, thanks not just to *Silent Spring* and *Population Bomb,* but to a myriad of well-timed ecocentric events.

---

23 Sarah Kate Tempelton,"Having Large Families is an Eco-Crime," *The Sunday Times,* May 6, 2007.
24 Natasha Courtenay-Smith and Morag Turner, "Meet the Women who Won't Have Babies Because They're not Eco-Friendly," *London Daily Mail,* November 21, 2007.
25 Gregory M. Lamb, "Earth's big problem: Too many people," *Christian Science Monitor,* January 28, 2009.
26 Ehrlich and Holdren, "Population and panaceas, a technological perspective," *BioScience,* Vol. 19, No. 12, December, 1969.

1969—SUMMER OF LOVE, YEAR OF DECEIT

In January 1969, a Union Oil drilling platform six miles off the coast of Santa Barbara, California, sprang a leak, allowing hundreds of thousands of gallons of crude oil to seep into the Pacific and wash ashore. The cameras of the world's media rushed to the scene to focus on oil-coated birds stuck in the same muck that was used to power America's cars. The newly hatched ecology propagandists soaked it up.

The nation's first outspoken congressional environmentalist, Wisconsin Senator Gaylord Nelson, immediately flew to California to see the crisis for himself. An anticapitalism democrat, Nelson returned to Washington angered at the oil industry, vowing, "to do something to wake America up."[27]

Emotions still run high in ultraliberal Santa Barbara over the 1969 leak. Even today, as black, marble-sized balls of coagulated crude are often found interspersed on the beaches of Santa Barbara, deceptive local activists will direct naïve eyes toward the oil platforms offshore, fervently declaring that capitalism and big oil are to blame for the tar balls on their sand—but this observation is total fraud. There is so much oil just beneath the ocean floor off Santa Barbara, that the black gold is constantly seeping into the open waters at a rate of up to 170 barrels per day.[28] However, despite the vast supply of crude naturally bubbling up from the deep, all these years later the 1969 accident has fixed it in the minds of many that drilling for oil is dreadful, and we should feel guilty for using the gooey fossil fuel—period.

Six months later, in June, another event was etched into the American psyche. The Cuyahoga River in Cleveland, Ohio, became a symbol of a planet in disrepair when it caught fire . . . sort of. The popular memory is that the entire river was consumed in flames and burned for hours like a portent of the Bible's Book of Revelation. The fact is the fire burned for less than 30 minutes and no conclusive evidence of its cause has ever been determined, though it's widely accepted that the combination of

---

27 "History of Earth Day," Community Environmental Council, Santa Barbara, accessed January 28, 2007, http://www.communityenvironmentalcouncil.org/Events/Earthday/.
28 "Emissions Estimates for the Coal Oil Seep Field," UCSB Hydrocarbon Seeps Project, August, 2000, http://seeps.geol.ucsb.edu/.

industrial waste and floating debris somehow ignited beneath a train trestle. The entire blaze was extinguished so quickly that nary a photographer had time to snap a photo. The nightly news had to settle for film of a fireboat hosing down the charred span.

Nevertheless, images of the event are seared into the mind of Carol Browner, Obama's self-anointed "Climate Czar" and former director of the Environmental Protection Agency during the Clinton years. Browner once claimed, "I will never forget the photograph of the flames, fire, shooting right out of the water in downtown Cleveland. It was the summer of 1969, and the Cuyahoga River was burning."[29]

Quite a vivid recollection for a girl only 14 years old at the time. Her parents must have subscribed to *Time* magazine. A week after the fire, *Time* ran a dramatic cover shot of the burning Cuyahoga River—from a much more serious fire in 1952![30]

The relentless coverage of America's pollution issues in '69 even made it to the floor of the United Nations, with Secretary General U Thant predicting that the planet had only ten years to avert environmental disaster. Naturally Thant, an admitted socialist (he said he did not like communism's "violent" tactics[31]) blamed the bulk of the coming planetary catastrophe on the United States.[32]

Incredibly, only twenty-four short years after sparing the world from speaking German, the great United States was now vilified as the earth's problem: its people and corporations were trashing the planet with pollution. In addition, there was an added wrinkle of angst festering amongst America's critics, both foreign and domestic: its military machine was involved in a costly and unpopular war against communist forces in Southeast Asia.

Fed up, bummed out, searching for answers and seeking a collective coming together, in August, 1969, 200,000 antiestablishment, young hippies trekked to a farm in upstate New York for the three-day Woodstock Music and Art Fair. The event, which the hippies referred to as being the climax to the "summer of love," represented the largest rock concert to

29 Jonathan Adler, "Smoking Out the Cuyahoga Fire Fable," *National Review*, June 22, 2004.
30 Ibid.
31 "Neutralist with Moral Fiber," *TIME*, November 10, 1961.
32 Jack Lewis, "Birth of the EPA," *EPA Journal*, November, 1985.

date, with most of the musicians utilizing the stage to fuel the fires of rebellious change. Millions who could not attend purchased the ensuing soundtrack and viewed the concert's documentary in theatres across the country. Woodstock represented a new ethos defined by sex, dope, rock and roll, and a loathing of mom and dad's America.

Empowered by the pharmaceutically enhanced high of Woodstock, that autumn nearly half-a-million people participated in antiwar demonstrations in Washington, D.C., with similar demonstrations held in San Francisco and throughout the United States on college campuses large and small. The Land of Liberty had never experienced such domestic unrest, and the seditious leaders of this burgeoning movement believed, if sustained, this home-grown displeasure could eventually transform policy on a variety of Marx-inspired levels. Cleverly realizing the war protests were a moment in time, their focus would soon return to "saving the planet from pollution"— a plot that would provide their lot with long-term momentum.

Their next step would include tearing a page from the playbook of Vladimir Lenin: they would take over the classrooms.

IN HONOR OF COMRADE LENIN

Senator Nelson was at the forefront of radical social engineering, and he was using my generation as his guinea pig. Nelson had created a protest tactic that was picking up steam: the "teach-in." During these teach-ins, mutinous school instructors would scrap the day's assigned curriculum, have their students sit cross-legged on the floor, and "rap" about how America was an imperialist nation and why communism really wasn't such a bad form of government—it just needed to be implemented properly.

His efforts were being aided by a young man named Denis Hayes, a former student body president at Stanford with an effective track record for organizing antiwar protests. While pursuing a master's degree in public policy at Harvard, Hayes had heard about the teach-in concept and sought out Nelson to help him take his strategy of infiltrating the classroom nationwide.[33]

---

33 Richard Seven, "Treading Lightly One Small Step at a Time," *Seattle Times*, April 21, 2002.

By the fall of '69 teach-ins were popping up all over the country and were surprisingly being tolerated by school officials.

It was about this time that Nelson met with Paul Ehrlich to discuss, we are told, overpopulation,[34] but the conversation apparently involved utilizing the pollution scare to achieve their long-term, seditious goals.

"My God," Nelson said, following his meeting with Ehrlich, "why not a national teach-in on the environment?"[35]

Years later, Nelson elaborated, "I was satisfied that if we could tap into the environmental concerns of the general public and infuse the student antiwar energy into the environmental cause, we could generate a demonstration that would force this issue onto the political agenda."[36]

Soon Senator Nelson formally announced there would be a "national environmental teach-in" sometime in the spring of 1970.[37] He would utilize surrogates like Ehrlich to prime the pump. Hayes would play a pivotal role in the organization and implementation. After careful consideration a name and date for the event were chosen: Earth Day would be celebrated each April 22.

Skeptical historians immediately noted a bizarre coincidence. The date coincided with the 100th anniversary of the birth of Lenin. Earth Day organizers have since tried to brush aside the date with lame retorts like, "Lenin wasn't an environmentalist." But he didn't have to be. Lenin's core political philosophy was linked at the hip with these newly fangled environmentalists who maintained that America's government must be altered, its economy planned and regulated, and its citizens better controlled. The environment would be the perfect tool to force these changes, and the most efficient way to gain converts would be through the public school system—the earlier the better.

Nelson and Ehrlich were already known as nontraditional crackpots,

---

34 History 179, North American Environmental History, Lecture 20, Brown University, http://www.brown.edu/Courses/HI0179/Lectures/Lecture_Twentyone.htm.
35 Ibid.
36 Senator Gaylord Nelson, "How the First Earth Day Came About," *American Heritage Magazine*, October, 1993, http://earthday.envirolink.org/history.html.
37 Tim Brown, "Earth Day and the Rise of Environmental Consciousness," United States International Information Programs, State Department, April 11, 2005, httpi//usinfo.state .gov/gi/Archive/2005/Apr/11-390328.html.

but young Hayes was that and more. In a *New York Times* article published the day after the first Earth Day entitled, "Angry Coordinator of Earth Day," young Hayes bragged that five years earlier he fled overseas because "I had to get away from America." Hayes was so committed to his anti-capitalist cause that he made sure his organization did not even produce any Earth Day bumper stickers. "You want to know why?" he explained to the *Times,* because "they go on automobiles."

Hayes had obviously returned to America, ready to invigorate the vision of comrades who had gone before him: comrades like Vladimir Lenin, who believed the schools would be their best means of indoctrination. "Give me four years to teach the children," Lenin is famously quoted as saying, "and the seed I have sown will never be uprooted."

The one-year anniversary of the Santa Barbara oil leak was used as a prequel to the inaugural Earth Day. A thousand people gathered at Santa Barbara City College to hear Paul Ehrlich, who no doubt reflected that the spill validated the claims made in his book, and Denis Hayes, who most certainly whipped the crowd into a frenzy, informing them that in just three short months they would be changing the world with the observance of a new "holiday"; one that would even be celebrated in the classrooms of America.

Looking back, if I would have known then what I know now, I would have ditched science class on April 22, 1970, and my dad would have backed me all the way to the principal's office.

FROM EHRLICH TO OBAMA

Earth Day has never been a celebration of God's wonderful creation; instead it's always been an assault on man. "Man must stop pollution and conserve his resources," championed the *New York Times* in an April 23, 1970, editorial, "not merely to enhance existence but to save the race from intolerable deterioration and possible extinction."

During that first Earth Day man was proclaimed the polluter and would remain as such for subsequent observances that decade. By the Eighties the event's organizers cast man as the tree killer, and, with the Nineties, man evolved into the animal species annihilator. The global warming scare never really became popular until the late Nineties, and when it did, it provided a hook that the compatriots at the Earth Day headquarters could hang their red berets on. Known as anthropogenic global warming, it was a sexy sell: humans—particularly Americans—were now screwing up the entire planet's *weather.* By 2000, Earth Day organizers took ownership of this new angle and would never let go.

Interestingly, in 1990, Paul Ehrlich would retread old ground with a new release, *Population Explosion.* In this popular treatise, the fiery Ehrlich blames virtually every human catastrophe, both real and imagined, on overpopulation (of course), and religion—especially the Catholic Church. And guess who wrote Ehrlich's dust jacket endorsement? None other than a Tennessee senator named Al Gore: "If every candidate for office were to read and understand this book, we would all live in a more peaceful, sane, and secure world."

In 1994, Ehrlich presented a high-level speech to the United Nation's International Conference on Population and Development held in Cairo, Egypt. In his address entitled, "Too Many Rich People," he quotes new thoughts from his writing pal John Holdren, envisioning a complex future energy cap and trade program by which "the gap between rich and poor nations would be closed."[38]

Al Gore would go on to become vice president, with Ehrlich empowered as a trusted advisor. Gore would create a movie about anthropogenic global warming called, *An Inconvenient Truth.* Despite a script packed with fraud, the film would receive an Oscar. Gore would continue to elude debate on the topic of global warming and still manage to be awarded a Nobel Peace prize for his "work," using the prize money to fund his PR firm, the Alliance for Climate Protection, located just down the street from Professor Ehrlich's office on the Stanford campus. Gore would lend fatherly advice on global warming to President Barack Obama, and Obama would hire Ehrlich's pal John Holdren to be his trusted science advisor. Obama would push an updated version of a cap and trade scheme originally envisioned by Holdren many years before, while Gore,

---

38 Paul Ehrlich, "Too Many Rich People: Weighing Relative Burdens on the Planet" speech presented at the United Nations International Conference on Population and Development, Cairo, Egypt, September 5-13, 1994.

now a member of one of Silicon Valley's most prestigious venture capital firms (and conveniently located just down the street from his Climate Protection office), would be comfortably in place to make billions off the conspiracy.

CLIMATEGATE

Though the following chapters will disclose multiple layers of deep climate fraud, providing you with the confidence to articulate your beliefs in the face of dogmatic opposition, the most damning crack in the foundation of anthropogenic global warming was discovered in November of 2009. Over a thousand emails, leaked from an internal computer system within the Climate Research Unit at the University of East Anglia in the United Kingdom, reveal how a small group of highly influential British and U.S. scientists have for years been secretly discussing ways in which their research could be maneuvered to make their case for human-caused climate change.

The Climate Research Unit (CRU) had been regarded by many as one of the most credible atmospheric institutions in the world, but with the revelation of the email exchanges, their credibility was reduced to junk science. The emails reveal that the world's leading climate scientists were working together to block Freedom of Information requests to review their data; marginalize dissenting scientists; manipulate the peer-review process; and obscure, massage, or delete inconvenient temperature readings. One certainly wonders, why? Especially since Al Gore has assured the world that "the science is settled."[39]

One answer seems to be simple: just follow the money. After all, that model seems to work pretty well for the former vice president.

Consider Phil Jones, the director of CRU and Climategate's key player. According to one of the leaked documents, between 2000 and 2006 Jones was the co-recipient of roughly $19 million worth of research grants, six times what he was awarded in the previous decade. It seems the louder

---

39 Andrea Seabrook, "Gore Takes Global Warming Message To Congress," *NPR*, March 21, 2007, http://www.npr.org/templates/story/story.php?storyId=9047642.

Jones yelled "fire," the more the money poured in. As you'll discover in subsequent chapters, similar abuses of the First Amendment have been occurring within the hallowed halls of America's National Aeronautics and Space Administration (NASA) for years.

Another answer is that those involved likely cling to a Marxist political agenda.

As for Jones, as soon as the emails were revealed, he and his reckless colleagues were quick to turn the tables. "My colleagues and I accept that some of the published emails do not read well. I regret any upset or confusion caused as a result. Some were clearly written in the heat of the moment, others use colloquialisms frequently used between close colleagues," he said in a statement.[40]

Another co-conspirator whose missives were a part of the science fiction stew, Michael Mann (whose sloppy scientific methods will be exposed later), whined, "What they've done is search through stolen personal emails—confidential between colleagues who often speak in a language they understand and is often foreign to the outside world."[41]

Allow me to present a quick translation of Mann's confidential "language" so we can better understand this foreign tongue. "AR4" refers to the fourth assessment of the Intergovernmental Panel on Climate Change, released in 2007. AR4 is a masterfully slick, group-think document concocted by the United Nations to perpetuate claims of global warming. "Keith" likely refers to an important member of CRU whose last name is Briffa. Many believe Keith Briffa has created a boatload of documents that attempt to minimize past climate temperature fluctuations in order to paint the present as the warmest weather ever.

Now, that said, here's a staccato May 2008 exchange between Jones and Mann that implies volumes: "Mike, can you delete any emails you may have had with Keith re AR4?"

Hmmm. I know Mann claims this is a tough language to understand, but like you, I'm wondering why would the head of a premier climate research center ask Michael Mann—a guru in the world of global warming—to expunge from the record correspondence he's had with another climate

---

40 "Rigging A Climate Consensus," *Wall Street Journal*, November 27, 2009.
41 Ibid.

crony regarding the sacred IPCC report? Could it be because so many were attempting to use the U.K.'s Freedom of Information Act to take a look at CRU's data? Deleting anything implies a cover-up.

But of course, any challenges from critics and skeptics outside this cabal are quickly dismissed.

For example, in another leaked email, dated September 2009, Mann explains to a *New York Times* reporter that, "those such as McIntyre who operate almost entirely outside of this system are not to be trusted."

Again, as you will discover later, Stephen McIntyre is a brilliant researcher who has wholly discredited some of Mann's most significant work—work which has been heralded by the IPCC and Al Gore as gospel.

In a March 2003 email, Mann responds to a paper published in the journal *Climate Research*, calling for either a boycott or hostile takeover of the journal. Two months prior, in the January edition of the journal, a short paper was presented by Harvard-Smithsonian astrophysicists, Sallie Baliunas and Willie Soon. They had reviewed more than 200 climate studies which determined that the twentieth century was neither the warmest century, nor the century with the most extreme weather during the past 1000 years. Snubbed, Mann wrote, "Perhaps we should encourage our colleagues in the climate research community to no longer submit to, or cite papers in, this journal. We would also need to consider what we tell or request of our more reasonable colleagues who currently sit on the editorial board."

In an especially flagrant email to Phil Jones, Mann brazenly reveals his desire to manipulate data, rip fellow climate scientists, and impose outright censorship:

> The attachment is a very good paper—I've been pushing Adrian over the last weeks to get it submitted. . . . The basic message is clear—you have to put enough surface and sonde obs [in other words, *add more* surface and air temperature readings] into a model to produce Reanalyses [obtain the result you wish to see]. The jumps [increases in temperature] when the data input change stand out so clearly. NCEP [the forecast methods used by the National Centers for Environmental Prediction] does many odd things also around sea ice and over snow and ice.

The other paper by MM [Stephen McIntyre and Ross McKitrick] is just garbage—as you knew. [Dr. Chris] De Freitas again. [Dr. Roger] Pielke is also losing all credibility as well by replying to the mad Finn [probably Finnish lawyer and vocal global warming skeptic Timo Hameranta] as well—frequently as I see it.

I can't see either of these papers being in the next IPCC report. Kevin and I will keep them out somehow—even if we have to redefine what the peer-review literature is!

Further tripe is revealed in another email from Mann to CRU's Tim Osborn. In an exchange dealing with temperature data going back to 1000 A.D., Mann is apparently troubled by the "redness," or warmth, of the data. Fearful that the information could get out to skeptics and deniers of global warming, Mann states:

> p.s. I know I probably don't need to mention this, but just to insure absolutely clarity on this, I'm providing these for your own personal use, since you're a trusted colleague. So please don't pass this along to others without checking w/ me first. This is the sort of "dirty laundry" one doesn't want to fall into the hands of those who might potentially try to distort things. . . .

Additional emails refer to the death of one prolific global warming denier as "cheering news," decry the work a well-known professor at MIT as "crap," while another wishes to "beat the crap" out of a noted skeptic.

My old man had a saying regarding guys like Mann, Jones, and their ilk: "if they want to fight dirty, beat 'em quickly, soundly, and with class." Thanks, Dad. Read on, my friends.

2

CLIMATE LOBOTOMY

This . . . was the most frightening aspect of the party regime—that it could obliterate memory, turn lies into Truth and alter the Past. The party slogan was "Who controls the past controls the future; who controls the present controls the past."

—George Orwell, *1984*

WHEN CITIZENS LACK a frame of reference, they are primed to succumb to the transformational vortex of historical revisionism. Karl Marx understood this well and wrote, "History does nothing; it possesses no immense wealth; it wages no battles."[42] Those waving the green flag of global warming respect history as little as Marx did.

No one living today was present to witness the first time the media, and a host of melodramatic scientists had gone gaga over global *cooling*. Lost in the dusty annals of history is the great Ice Age panic that spanned the nineteenth and early twentieth centuries.

On February 24, 1895, the *New York Times* started the ball rolling with the proclamation, "Geologists Think the World May Be Frozen Up Again." Due to concerns of rapidly advancing glaciers and reports of extremely cold weather around the world, the story wondered whether "recent and long-continued observations do not point to the advent of a second glacial period."

---

42 Karl Marx, *The Holy Family* (1846), ch. 6.

"Glacial period" meant Ice Age, and for the *Times* it meant newspapers selling like hotcakes—readers have always loved a colossal scare.

Predictions of a coming global freeze escalated when the much publicized *unsinkable* Titanic met its fate with an uncharted iceberg in the North Atlantic on April 14, 1912, killing some 1,500 people, including many of the world's rich and famous. Three months after the tragedy, page one of the *Times* heralded, "Prof. Schmidt Warns Us of an Encroaching Ice Age." In that October 7, 1912, story, the *Times* quotes Professor Nathaniel Schmidt of Cornell University as saying the world will need scientific knowledge "to combat the perils" of the next ice age. Unbeknownst to Dr. Schmidt during his day, modern-day global warming hucksters would use the exact same lingo to panic their audiences ninety years later. By 1930, the cooling hysteria had done nothing more than sell a lot of ad space, as temperatures began climbing during the hot and dusty Thirties, setting records that, in many cases, have yet to be broken.

The problem with every generation is that a long-term memory of the past requires a determined and studied effort—a fact upon which modern eco-Marxists depend for success. In this age of information/false information overload, even the recent past quickly becomes fuzzy, almost guaranteeing a headache to anyone who racks his brain digging deeply to mine true facts and details. And that's precisely what the current climate-scare tacticians are banking on, especially the elitist politicians and policymakers who continually capitalize on a society's lack of cognizance—the bigger the public's memory hole, the better. They see it as an effective means to grow government and thus, better control the way in which the underclass lives. In addition to the deceivers in government, unscrupulous scientific researchers are dependent on perpetuating the global warming and climate change myth for continued flows of grant money. Colleges and universities have established curricula and majors to indoctrinate an entire generation to enlist in the so-called "green workforce." Without a crisis, many politicians, researchers, educators, and graduates are without a cause.

However, contrary to Marx, history *does* produce great wealth and *can* win battles—especially in the climate debate—because, *climate happens*, without the influence of man.

LITTLE ICE AGE

From roughly 1350 to 1800, the earth's temperature was undeniably colder than today. Given that the thermometer was not mass-produced and in widespread use until the 1800s, the conclusions we draw regarding this 500-year period are based on physical evidence culled by honest researchers and plentiful historical observations that clearly illustrate the period known as the Little Ice Age (LIA). The evidence includes hundreds of studies illustrating the physical isotopes of carbon, hydrogen, and oxygen in decayed plants excavated throughout Europe, indicating significantly lower temperatures during the LIA.[43] The examination of tree rings using boreholes indicates a colder climate worldwide (a tiny hole is bored into the cross section of a tree trunk with the long, tubular contents removed and closely examined; the number of rings corresponds to the number of growing seasons, with the width of the ring illustrating temperature conditions during a particular year).[44]

Studies of coral, which exhibit their own annual growth bands, yields data that suggests average ocean water temperatures cooled significantly during the LIA.[45] Aggressive studies on past climate instituted by independent Chinese researchers unanimously conclude, "China has advantages in reconstructing historical climate change for its abundant documented historical records and other natural evidence obtained from tree rings, lake sediments, ice cores, and stalagmites."[46] The brilliant collective research by the Chinese points to the LIA being nearly 2°F (1.2°C)[47] cooler than today,[48] which is in line with other global calculations from the period. Studies from New Zealand,[49] Chile,[50] and

---

43 T. Richard D. Tkachuck, "The Little Ice Age," *Origins* 10, no. 2 (1983), 51–65.
44 Ibid.
45 Druffel (1982).
46 Ge et al. (2004).
47 Henceforth, whenever I display the temperature in both Fahrenheit and Celsius, it is the Celsius figure that is the official temperature of record.
48 Ibid.
49 Williams et al. (2004).
50 Koch and Kilian (2005).

Africa[51] all point to unified conclusions: the Little Ice Age was a global event experienced worldwide.

If you are new to the details of the global warming debate, it is imperative to understand that the above citations are critical (and only scratch the surface of a wealth of supporting evidence) because global warming activists stubbornly maintain that the Little Ice Age was a local event which exclusively impacted isolated sectors of the northern hemisphere. Obviously, that's a lie.

Complementing the physical evidence are dramatic historical accounts from this same period. Not only are these corroborating narratives incredibly compelling, they should cause us to pray that today's comfortable climate continues because, in all actuality, the past climate was miserable.

FROZEN PILGRIMS

History books have long ago erased the tenacity of America's first Pilgrims, whose three-month journey across the Atlantic aboard the *Mayflower* brought them to an unbearably cold corner of North America, which we now know as Cape Cod, Massachusetts.

Prior to their journey, the Pilgrims were well-acquainted with frigid weather. Summers back home in England were rarely warm, and snow, wind, and bitter cold were a normal part of their long winters. In fact, London's Thames River froze solid nearly every year—often with ice a foot thick (in contrast, the river only froze four times in the twentieth century and *never* a foot thick). Once the ice was deemed safe to walk on, the entire city would be drawn to the icy shores. Frost Fairs would convene on the middle of the river, complete with merry-go-rounds, swings, slippery soccer matches, and chaotic donkey and horse races. Hooligans and rabble-rousers would also take advantage of the phenomenon, turning the Thames into a frozen river of sinful delight. Diarist John Evelyn described activities of the Frost Fairs as ". . . sleds, sliding with skates,

---

51 P. deMenocal et al. "Coherent High- and Low-Latitude Climate Variability During the Holocene Warm Period," *Science*, v. 288, p. 2198-2202, June 23, 2000.

bull-baiting, horse and coach races, puppet plays and interludes, cooks, tippling [taverns for the single purpose of getting wasted] and other lewd places, so that it seemed to be a bacchanalian triumph, or carnival on the water."[52]

It was September 1620, when the Pilgrims left their English homeland and, upon arrival at Cape Cod in November, were about to be greeted by the full fury of the Little Ice Age, North American style.

Writing in his *History of the Plymouth Settlement, Mayflower*, Pilgrim leader and governor of the Plymouth Colony, William Bradford, wrote that within weeks of their arrival in the New Land, "the severity of the winter weather, and sickness, had begun. . . ." Over the next three months nearly half of their company died, "partly owing to the severity of the winter, especially during January and February," Bradford reported. "Of all the hundred-odd persons, scarcely fifty remained, and sometimes two or three persons died in a day. In the time of worst distress, there were but six or seven sound persons who, to their great commendation be it spoken, spared no pains night or day, but with great toil and at the risk of their own health, fetched wood, made fires . . . and showed their love to their friends and brethren."

Were it not for the partnership forged with the local Indians, and the essential provisions these natives were able to supply to their new Pilgrim friends, it is unlikely that any of the English sojourners would have survived through spring—and their deaths would not have been an isolated incident. Climate-caused mortality was widespread during the seventeenth century. One of the worst episodes occurred in Europe due to the failed harvest of 1693. The shortage of food became so severe that livestock was slaughtered for lack of feed and *millions* of people died of starvation.

The Little Ice Age prolonged through the eighteenth century and almost stole America's independence in 1777, as George Washington's crucial stand against the Brits at Valley Forge nearly ended in defeat because hundreds of his men succumbed to severe frostbite.

Thankfully, by about 1850, the climate of the earth began to noticeably

---

52 J.B. Priestley, *The Prince of Pleasure and His Regency* (London: William Heinemann Ltd., 1969), p. 113.

warm. Five-hundred frigid years of human history and weather-induced misery had concluded.

However, these facts are generally ignored by agenda-driven researchers guilty of purposely concealing this historical climate record and coercing the naïve to believe earth's temperature has always been relatively constant . . . until now. And those who dare speak otherwise are *deniers.*

"YOU DON'T KNOW WHICH FACTS TO LEAVE OUT"

I first received the title of "denier" in 1997. I had been called from my weather office into the newsroom by a hyped-up producer who was buzzing about the newest fad, *global warming.* Vice President Al Gore was coming to California to convene a "summit" to discuss the current El Niño weather patterns we were experiencing in the western states, and their relationship to global warming.

"Unreal," I thought as he described his plan for our televised team-coverage. "The newsroom is about to link a perfectly natural, periodic weather pattern with global warming. This stinks."

They wanted me to front the evening's coverage of the coming summit.

"But there's no scientific evidence linking global warming to El Niño," I explained.

The producer rolled his eyes, slowly wagged his head and glared at me.

"Brian," he condescendingly replied, as if correcting a confused child, "people are surfing without wetsuits in San Francisco. Temperatures are up all over the world. Sea levels are rising. Glaciers are melting. What are you talking about, 'no evidence?' It's the biggest story in years. You're in denial!"

Denial? Nearly two decades in the weather business, a shelf full of awards, a "Brian Sussman Day" proclaimed by the State of California . . . forget denial, now I was ticked! However, I knew the science, and I knew the meteorological history, and I knew I could not just let this moment pass. Good old Dad was welling up inside me.

"Let's get this straight," I began. "El Niño is real, but global warming is a total farce!"

I had an easy-going reputation among colleagues and was not known for making a scene, so heads around the newsroom snapped to attention at my comment.

Now or never, I thought, and like an uncapped fount of knowledge, I unloaded.

"You want denial? Look in the mirror. During the last El Niño in 1982 schools of marlin from L.A. waters migrated 400 miles north to our coast. No one was talking about global warming then. Do you recall that? Of course not, it doesn't fit your story!"

The newsroom went silent. I'd crossed a line.

"And FYI, the earth was much warmer in the past than it is today—you ever heard of the Vikings in Greenland?"

Of course he had. The producer glared. My verbal blows were landing.

"And your sea levels? They're not alarming. And your melting glaciers? For crying out loud, I took a bunch of our viewers to Alaska a few years ago, and we saw glaciers that were *growing*, not melting. You're not just in denial—you've got an agenda!"

Awkwardly aware of our newsroom audience, the producer turned his back to our colleagues and stepped in closer.

"Brian," he seethed, lowering his voice, "your problem is simple. You don't know *which facts to leave out*."

I was floored. Aware of the editorial bias that viewers were often subjected to, I never heard it stated so succinctly. There was no reason to stretch our argument out any further. Despite years of hard work to become a reliable and trusted source of weather information for the six million people in our audience, inside the walls of our TV station I was now a "denier."

The cover story aired, but without me as the front man. Instead, I presented the viewers with yet another primer on El Niño and its potential effects on both California and the rest of the country. I made no mention of global warming—I didn't have to—everyone else in our team-coverage did.

MEDIEVAL WARM PERIOD

During my newsroom argument, I had blurted to the producer, "the Vikings in Greenland." This is a sore subject to promoters of global warming because, without modern conveniences, life in Greenland today would be impossible. The Vikings lived during the "Medieval Warm Period"—a four hundred year epoch that ruins the global whiners' claims that today's climate is the warmest *ever.*

The Medieval Warm Period (MWP) occurred from 900 to 1300 A.D. During this span the overall average mean global temperature was considerably warmer than present. There were no SUVs, smokestacks, or airplanes seeping the currently vilified, ever so minor greenhouse gas, carbon dioxide into the atmosphere. The MWP was a dramatic climatic event that *just happened.*

Again, as with the Little Ice Age, we can confidently make our case based on an abundance of physical and historical evidence, including the world-famous wines once produced in, of all places, the British Isles.

In recent centuries Great Britain has been well-known for hearty beer and robust whiskey, but certainly not wine. However, during the height of the MWP, England was home to some of the world's finest commercial vineyards. The grapes cultivated and wines produced in England, from 1100 to 1300, competed well with those in France and Germany—regions that continue to produce quality wine grapes today. The important item to note is that the former vineyards in England are located latitudinally several hundred miles *north* of those currently found in France and Germany and are completely incapable of yielding wine grapes today— the climate is simply too cool.

Looking southward, the modern vineyards of Germany provide further evidence of a past warmer climate. During the MWP, German vintners were growing grapes at significantly higher elevations than they can today. Examining the high slopes above their current wine country, remnants of the ancient vineyards remain visible at the 2,500-foot elevation. Currently, Germany's favored wine grapes are grown below 1,500 feet, with a select few fields doing well within unique microclimates reaching almost 1,800 feet. Utilizing a bit of simple math based on something we refer to in the world of atmospheric science as the adiabatic lapse rate,[53] the difference in average temperature back when grapes could thrive

---

53 Based on international standard atmospheric conditions, temperatures will change at a rate of 6.49°C per 1,000 meters (3.56°F or 1.98°C/1,000 ft.).

at 2,500 feet, versus today when the fruit generally does not grow well above 1,800 feet, indicates that the climate there was at approximately 2°F (1–1.4°Celcius) *warmer* than today.

Further botanical evidence from Europe also places another nail in the whiner's coffin. By observing the Alps, one can clearly see the remnants of ancient trees that once thrived at as much as 1,500 feet above the current tree line. Similarly, present tree lines in Iceland also indicate that the forest has receded about the same distance. In fact, occasionally, entire, mature, ancient birch tree trunks are spit out from the end points of massive calving Icelandic glaciers, indicating that there is a frozen forest beneath the sluggish rivers of ice and snow.

And there is ample proof that the MWP, like the Little Ice Age, was a global climatic event: in the Atlantic Ocean's famed Bermuda Triangle, radiocarbon dating of marine organisms in sea bed sediments illustrate that sea surface temperatures were nearly 2°F (1°C) warmer during the MWP than they are today;[54] at Kenya's Lake Naivasha, extracted sediments from the lake bed reveal the lake endured a lengthy 200-year drought from about 1000 to 1200 A.D.;[55] in Peru, ice core samples from the Quelccaya Glacier in the Andes demonstrate that average temperatures during the MWP at times spiked above the readings of today;[56] in a study of oxygen isotopes in a peat bog in northeastern China near the border of North Korea, Taiwanese researchers uncovered a 6,000-year temperature history, illustrating that the temperature between 1100 and 1200 A.D. was nearly 2°F (1°C) warmer than it is today; they also found extremely cold temperatures between 1550 and 1750, in conjunction with the Little Ice Age.[57]

Again, like the evidence supporting the global effects of the Little Ice Age, this is a thumbnail of the multitude of research supporting this position. However, the most entertaining example of an entire culture that thrived due to the Medieval Warm Period involves the wild and crazy Vikings.

---

54 L.D. Keigwin, "The Little Ice Age and Medieval Warm Period in the Sargasso Sea," *Science*, v. 274, Number 5292, 1996 pp. 1503-1508.

55 D. Verschuren, "Rainfall and Drought in Equatorial East Africa during the past 1,100 Years," *Nature*, v. 403 (6768), January 27, 2000 pp. 410-414.

56 http://academic.emporia.edu/aberjame/ice/lec19/fig19d.htm.

57 Hong et al., "Response of Climate to Solar Forcing," *The Holocene* 10 (2000), 1-7.

VIKINGS' BIG BREAK

Eric Thorvalsson was born in Norway in 950 A.D., the dawn of the Norse golden years. The climate had substantially warmed, and now these simple Scandinavians, known in the past for being meager fur traders, were sowing their oats: donning metal hats with protruding cattle horns, building ornate ships that appeared to have been designed while under the effect of hallucinogens, and sailing to other lands to go "a-Viking" (i.e. pillage and plunder). Perhaps it was because the warmer climate allowed the Vikings to grow their own barley and hops to produce beer—which, for the first time, they did—but this new breed of Norsemen was fired up to conquer the world.

Erik was a red-haired hothead, and so nicknamed, Erik the Red. His father had been banished from Norway to Iceland for murder, and following in pop's footsteps, young Erik ended up killing two men in Iceland and was expelled for a three-year period beginning in 982. Erik assembled a crew and sailed west in exile, arriving on a nice chunk of real estate he initially referred to as Erik's Island. His sentence expired, he returned to Iceland, convincing perhaps as many as 500 people to follow him back to his personal island, which, for promotional purposes, he accurately dubbed, "Greenland." For the next three hundred years, Greenland was a strategic outpost for the powerful Vikings, with its coastal communities developing into thriving whaling and fishing enterprises, while inland, farmers prospered, too.

Excavation of Viking villages and burial grounds have been uncovered beneath what is presently permanently frozen soil, suggesting to researchers that temperatures were perhaps as much as an astounding 7°F (4°C) warmer in that region than today.[58] Careful soil excavation has also shown the presence of corn pollen—corn cultivation is impossible in Greenland today. More astounding evidence illustrating a significantly warmer climate has been located north of Greenland on Ellesmere Island and the New Siberia Islands. In both of these remote locations, former homes

---

58 F. Donald Logan, *The Vikings in History* (London and New York: Hutchison and Company, Ltd, 1991), 78, cited by James R. Lee, *The Vikings In North America, Long-Term Climate Change, Environment, Trade, and Conflict*, ICE Case Studies, Number 86, June 2001, http://www1.american.edu/ted/ice/vineland.htm.

constructed of large timbers have been discovered in an area that is void of such tree growth—in fact the region is now an arctic desert.

In conjunction with the conclusion of the MWP around 1400, the Vikings' wild life of prosperity and plunder had ground to a halt. Their enemies had built effective defenses, the weather was returning to miserable, the growing sea-ice was inhibiting travel and Greenland was about to become inundated with a fresh round of glacier advances.

Certain whipper-snappers today would argue that the last remaining Viking ancestors can be seen on autumn Sundays afternoons in Minneapolis, wearing the same silly horn-hats, guzzling loads of beer, and swearing that Brett Favre *has to be* an ancestor of Erik the Red.

UNITED NATIONS OF MARX

The global body committed to distributing the core tenants of Marx is the United Nations, and, predictably, their minions are at the heart of the climate cover-up.

In 1990, the United Nation's International Panel on Climate Change (IPCC) presented the world with the following simple chart (Figure 2.1) noting temperature change since 900 A.D.

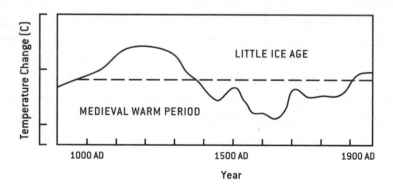

Figure 2.1 Medieval Warm and Cold Periods
Source: 1990 IPCC Assessment Report[59]

---

59 1990 IPCC Assessment Report graph, "Kyoto by Degrees," *Wall Street Journal,* June 21, 2005, A16.

The uncomplicated graph clearly illustrates the years comprising the Medieval Warm Period were significantly warmer than today, and the Little Ice Age similarly cooler. As of 1990, these two major climate shifts had never been disputed within the scientific community.

However, thanks in large part to Al Gore's vigilant propaganda campaign, which was unleashed in 1992 with his book *Earth in the Balance* and his election as Bill Clinton's vice president, more and more activists—some with science degrees—comprehended that such a graph was a campaign killer. During the MWP there clearly was no human-caused greenhouse gas connection. These activists feared that the public would never buy into the theory that greenhouse gases—especially the carbon dioxide (CO_2) generated from fossil fuels—were currently warming the planet if, during the warmest period on record, CO_2 was *not* a factor. Such a graph might provide the American masses the rationale to continue fueling their big cars and SUVs, heating and cooling their large homes, and maintaining their sights on progressing up the ladder of material affluence; all with no concern for the down-trodden third worlders. It was an injustice that could not be tolerated by Marxian minds.

To properly hawk modern-day global warming, a revisionist lobotomy needed to be performed on the record books; the question was how to accomplish this devious plan.

Enter a Ph.D. from the University of Oklahoma's College of Geosciences, David Deming. In 1995, Deming had concluded extensive climate research on ancient tree rings. Boring minute holes into the trunks of the trees, Deming examined growth patterns relative to past weather and climate.

I have interviewed David Deming and would describe him as both a gentleman and principled scientist—a real class act. If your son or daughter took a science course from him at the University of Oklahoma, I would guarantee there would be no opining from the lectern regarding social or economic justice. Deming would teach your children science—period.

Deming's jaw-dropping story is best told in his own words:[60]

---

60 David Deming, "Global Warming, the Politicization of Science, and Michael Crichton's 'State of Fear,'" *Journal of Scientific Exploration*, v. 19, no. 2, June, 2005, excerpted from a reprint available at http://www.sepp.org/Archive/NewSEPP/StateFear-Deming.htm.

In 1995, I had a short paper published in the prestigious journal *Science* (Deming, 1995). I reviewed how borehole temperature data recorded a warming of about one degree Celsius in North America over the last 100 to 150 years. I closed the manuscript with what seemed to me to be a remarkably innocuous and uncontroversial statement: "A cause and effect relationship between anthropogenic activities and climatic warming cannot be demonstrated unambiguously at the present time" (Deming, 1995, p. 1577).

The week the article appeared, I came into my office one morning to find an email message from a reporter for National Public Radio. He wanted to interview me concerning my article in *Science.* Visions of glory danced in front of my eyes. I was going to be on national radio. Surely, it was only a matter of time before I would be a regular guest on the McNeil-Lehrer news hour on PBS. Excited, I called the reporter back. But all of my fantasies were immediately dispelled. The reporter focused in on the last sentence in the *Science* paper. He asked me if I really meant to say that. Did I really intend to imply that the warming in North America might have been due to natural variability? Without hesitation, I said yes.

He replied, "Well, then, I guess we have no story. That's not what people are interested in. People are only interested if the warming is due to human activities. Goodbye."

And he hung up on me. It was my first realization that the media intentionally filter the information the public receives. . . .

I had been naïve. I had assumed that everyone was like me—that they were interested in the truth. But a lawyer's job isn't to discover truth; it's to win an argument. Neither is an advocacy organization interested in truth— they are committed to advocating a certain position regardless of the facts.

Despite National Public Radio's probing question and subsequent decision to blow off Deming's research, with the publication of his paper in *Science,* Deming gained significant credibility within the community of scientists working on climate change—they all figured he was in their camp. Deming continues:

They thought I was one of them, someone who would pervert science in the service of social and political causes. So one of them let his guard down. A major person working in the area of climate change and global warming sent me an astonishing email that said, "We have to get rid of the Medieval Warm Period."

The astonishing email Dr. Deming received was actually the first wave of the Climategate scandal. Since then, Deming has shared the account before a Senate committee in Washington, D.C., accurately referring to it as "historical revisionism." But I wanted the rest of the story, specifically the name of the emailer who desired to subterfuge history.

My interview[61] with Dr. Deming was recorded as follows:

Sussman: Did you know the emailer?

Deming: No. You must understand that following an article being published in a journal like *Science*, it's quite common to receive emails from colleagues working in similar fields.

Sussman: So, you didn't know this guy?

Deming: No.

Sussman: Did his suggestion that somehow the record of the Medieval Warm Period had to be altered strike you as odd?

Deming: Yes. I really didn't think people would take this nonsense seriously.

Sussman: Nonsense? Is that how you saw anthropogenic global warming back then?

Deming: It wasn't as big an issue then. I'm a geologist. I am used to observing events over long periods of time. Everyone talks about the [computer] models. I understand that the atmosphere is a very complex system, and a system that is categorically impossible to replicate in terms of future predictions. So, I have to rely on what we know and what we don't know. That's why it's important to study past climate. In fact, I have something I call Deming's 10 Rules of Science. One of them is "You don't know what you don't know." What we do know is that the climate of the earth has undergone major changes. What we do know is that after the last ice age the temperature of Greenland increased by

---

61 Interview conducted February 19, 2008.

perhaps as much as 50 degrees in a period of perhaps 10 years. Why did that happen? We don't know—but it happened.

Sussman: Back to the email. Many of us have heard the rumor that it was Jonathan Overpeck, the NOAA scientist, who has been on a tear for years to rid the books of the Medieval Warm Period, but I've been unable to find a record of you publicly admitting as such. Was it Overpeck?

Deming: It's been many years, and I've long since deleted the email, but to the best of my recollection it was sent by an Overpeck.

Sent by an Overpeck. Dr. Deming claimed he was unable to recall the first name, and I didn't want to press this good man further. However, if it were Jonathan Overpeck, it would make total sense. Overpeck is a government apparatchik working for the National Oceanic and Atmospheric Administration (NOAA) and has made quite a name for himself speaking at conferences and writing research papers belittling those who disagree with the anthropogenic global warming hypothesis.

My suspicions about Jonathan Overpeck being the guy who contacted Deming were confirmed in a 2008 email unearthed in the CRU leak. During an exchange with Phil Jones and others, Jonathan Overpeck, clearly agitated all these years later by the comments made by Dr. Deming before the Senate (even though Deming *never* mentioned Overpeck by name), denies making the "we must get rid of the Medieval Warm Period" statement.

The email states:

> I googled "We have to get rid of the warm medieval period" and "Overpeck" and indeed, there is a person David Deming that attributes the quote to an email from me. He apparently did mention the quote (but I don't think me) in a Senate hearing. His "news" (often with attribution to me) appears to be getting widespread coverage on the internet. It is upsetting. I have no memory of emailing w/ him, nor any record of doing so (I need to do an exhaustive search I guess), nor any memory of him period. I assume it is possible that I emailed w/ him long ago, and that he's taking the quote out of context, since know I would never have said what he's saying I would have, at least in the context he is implying.
>
> Any idea what my reaction should be?

I would recommend Overpeck and his selective memory take an early retirement. Though he denies the statement made to Deming, I discovered an official 1998 government press release regarding the MWP, quoting Jonathan Overpeck. It seems crystal clear that by 1998 his mission to see the record revised was accomplished, as he declares, "the so-called Medieval Warm Period did not exist."[62]

Seems like somebody's pants are on fire.

And to whom shall we give credit for the climate lobotomy that supposedly killed the "so-called Medieval Warm Period"? Enter Doctor Leaky: Michael Mann.

MANN'S MEMORY-BUSTING HOCKEY STICK

Mann's surgical subterfuge was performed while working for the University of Massachusetts' Department of Geosciences. Mann cleverly "used" the highly respected tree ring research as his primary data to account for the medieval years.[63] For more recent centuries, Mann included other measurement from sea sediment, ice core samples, and oxygen isotopes—all sound, legitimate sources. However, Mann's process began waxing weird when he threw into the mix temperatures recorded in major cities that have undergone huge artificially-induced upward swings due to the heat-trapping influences of concrete, asphalt, and steel (known as the Urban Heat Island effect). Stranger still, conveniently *excluded* was the highly accurate temperature record of the earth produced by flawless satellite instrumentation that (as we will discuss in the next chapter), demonstrates only a miniscule warming since 1979. In addition, Mann did

---

62 "'Twentieth Century Global Warming Unprecedented,' NOAA Scientist Reports," NOAA Press Release, December 7, 1998, http://www.publicaffairs.noaa.gov/pr98/dec98/noaa98-88.html.

63 Mann apparently inputted data collected by Dr. Shaopeng Huang of the University of Michigan, who examined 6,000 tree-ring boreholes from around the world. Like Dr. Deming's research, Huang's indicated a pronounced global warming was evident in the medieval period, with temperatures significantly warmer than today. See: Huang, Pollack and Shen, "Late Quaternary Temperature Changes Seen in Worldwide Continental Heat Flow Measurements." Geophysical Research Letters 24: 1947–1950, 1997.

not discard wild temperature anomalies like the monster El Niño warming of 1982–1983 and 1997–1998.

In the end, Mann's results seemed to justify his means. Through a stupefying, number-crunching mathematical exercise, he meticulously manipulated the data collected through the diligent labor of the dedicated scientists who preceded him. The result was like an ice pick to the climate's prefrontal cortex: the past had been forgotten. Mann's work was presented to the world in a paper published in *Nature*[64] and in another published in *Geophysical Research Letters*.[65] For the global whiners, the papers were like manna from Heaven.

The final object of Mann's climatic memory-buster was a bogus graph that looked like a hockey stick positioned horizontally, with the blade protruding straight up. In fact, the chart became known as "Mann's Hockey Stick" and became a popular devious device to convince the uninformed observer that the earth was undergoing an unprecedented fever (Figure 2.2).

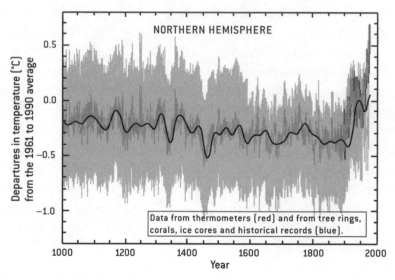

Figure 2.2 Mann's Hockey Stick
Source: International Panel on Climate Change Third Assessment Report, 2001.[66]

64 M.E. Mann, R.S. Bradley, and M.K. Hughes, "Global-Scale Temperature Patterns and Climate Forcing over the Past Six Centuries," *Nature*, 392, 1998, pp. 779-787.
65 M.E. Mann, et al., "Northern Hemisphere Temperatures During the Past Millennium: Inferences, Uncertainties, and Limitations," AGU GRL, v. 3.1, 1999.
66 IPCC Technical Summary, 2001, p. 29.

Observing Mann's stick, one can clearly see a dramatized temperature spike ensuing about 1930 and reaching warp-speed about 1970. Also, note that the Medieval Warm Period between 900 and 1300 and its counterpart, the Little Ice Age, have been expunged from the record.

Despite a horrendous bastardization of science, Mann's Hockey Stick has been accepted by millions. The United Nation still carries the Stick to compel global policy, while Al Gore carries it to the bank—the Stick is a pillar in his media presentations. And to this day, those who dare criticize the spurious Stick are branded "outliers."

However, proving that those like me—who see through their dangerous charades—are impossible to defeat in the arena of facts, yet another leaked email reveals further attempts to reshuffle the deck of history and force a more acceptable outcome. This communiqué was sent by Mann to Overpeck and a few others. Apparently, the plan was to reconstruct temperature data going back 2000 years, instead of the typical 1000, to accommodate their preferred revision of the MWP, even though they didn't have the necessary data:

> I think that trying to adopt a timeframe of 2K, rather than the usual 1K, addresses a good earlier point that Peck made w/ regard to the memo, that it would be nice to try to "contain" the putative "MWP," even if we don't yet have a hemispheric mean reconstruction available that far back.

This is not responsible science. If I may pull a quote from one of the leaked emails, it's cra—, (well, you get the idea).

SCIENTIFIC METHOD

Responsible science is built on skepticism; a lesson I learned the hard way. I was in eighth grade, had just moved to the area, and fellow students were already engaged in a science fair. I had no clue what was happening. All I knew was that by Monday I had to submit a project.

Looking back, I would have done just as well by jumping out of bed at the sound of my alarm Monday morn', throwing on my clothes, racing downstairs to the kitchen, grabbing a paper cup from the pantry, flying out the sliding glass door to Mom's garden, scooping up a single cup of dirt, and then sprinting to the bus stop. Once onboard the bus, I could rehearse my presentation: "Just add water and you get mud, teacher."[67]

I would have still received a lousy grade, but with a heck of a lot less stress.

I arrived at school to see a myriad of huge, elaborate, dad-made displays being hauled out the backs of station wagons. Though I was thoroughly embarrassed at my pitiful entry, the lesson I was about to learn would serve me well the rest of my life.

"What do you have here, Brian?" asked Mr. Phillips as he looked at my project.

"It's a perpetual motion machine." [*I am not making this up!*]

"Interesting," he said. "So, what was your observation?"

"Well, I was wondering if something could be set in motion forever."

"Have you ever heard of Newton's Laws of Motion?"

"Yes, sir."

"Give me the first one."

"Okay." I had this set to memory. "Every object in a state of uniform motion tends to remain in that state of motion unless an external force is applied to it."

"So, Mr. Sussman," he said, looking over the top of his black, horn-rimmed glasses, "you think you're going to defy Newton's first Law?"

"Well," I stammered, obviously trapped. "I think I was—I was—just trying to . . . show his law was real?"

"Okay, so really, this is an experiment designed to prove Newton's Law."

I was in over my head and confused.

"I don't know what it is," I confessed.

The teacher recognized my glazed look and had compassion on his new student.

"Brian, here's how the scientific method works. Look at this display."

We walked over to a table with a nice poster board, colorful graphics, and some plants.

"This girl is getting an A for her project. First, she made a simple

---

67 Thanks to comedian Brian Regan for inspiring this analogy. For a good laugh, buy his CD: *Brian Regan Live.*

observation: 'It seems like plants only grow where there is sunlight.' She then turned her observation into a question: 'I wonder if a bean plant can grow without sunlight?' Then she turned her question into a hypothesis: 'If a bean plant is not exposed to sunlight, it will not grow.'"

I was nodding my head up and down. It was all really quite simple.

"Now that she has a hypothesis," he continued, "she tests it, to prove it *wrong*. This is very important, Brian. You don't want to set up an experiment designed to positively prove your hypothesis. You want to rigorously attack the hypothesis," he said with passion.

Then this brilliant instructor said words I've never forgotten.

"If your hypothesis can be disproved, you must throw it out and come up with a new one."

"So, I see how she attacked her hypothesis," I said, as I observed the plant display. "She grew this bean plant on her window sill where there was plenty of sunlight."

"Yes, but first notice a few things."

The teacher illustrated that hers was a wonderfully controlled experiment. The student planted six identical bean seeds into six identical pots, all containing exactly the same amount of soil. The seeds were even planted at the same depth in the soil. Each seed was watered with the same amount of water. They were all raised in her house where the temperature was relatively uniform. The only variable for each seed was sunlight.

"Ah, the amount of sunlight—just what she was testing," I said.

"You're getting it, Brian," the judge said, genuinely excited. "These two pots were placed on a counter next to a kitchen window with plenty of sunlight, and we see two healthy bean plants. These two were placed next to the window with the curtains drawn. They were only able to receive filtered sunlight. Look at the plants."

They were both rather scrawny.

"Now, observe these pots," he continued. "They were placed next to a window blocked off with black construction paper. The bean seeds never saw a ray of sunshine."

The pots were void of life.

"Why two pots for each location?" I asked.

"The more test subjects you have for an experiment, the more valid your results. What if she had a bad seed? The experiment would be ruined. Now she has a nice conclusion, 'a bean seed needs sunlight in order to grow and will grow best with direct sunlight.' In fact, now we could say her hypothesis has moved up to becoming a scientific theory."

"What happens with a theory?" I asked.

"You continue to attack it to determine if there is any way in which you can cause it to fail."

"And then?"

"And then, if it can withstand all attacks, you have a Law—as in Sir Isaac Newton's Laws of Motion."

Grade-wise, I bombed out in the fair, but I scored well with a tremendous basic science lesson. In hindsight, it makes me wonder . . . where were Mann, Jones, Overpeck, Gore, and a host of other climate clowns during their science fairs?

BREAKING THE STICK

According to the National Academy of Science:

The fallibility of methods is a valuable reminder of the importance of skepticism in science. Scientific knowledge and scientific methods, whether old or new, must be continually scrutinized for possible errors. Such skepticism can conflict with other important features of science, such as the need for creativity and for conviction in arguing a given position.[68]

The initial chief skeptics of Mann's work were Stephen McIntyre and Ross McKitrick. McIntyre entered the fray as a curious mathematician, McKitrick as an equally interested associate professor in the Department of Economics at the University of Guelph, Ontario. Curious to see how the hockey stick graph had been created, in the spring of 2003, McIntyre and McKitrick contacted Mann, requesting his raw data. Such requests are quite common in the world of scientific research and clearly are welcomed

---

68 "On Being a Scientist: Responsible Conduct in Research," National Academy of Science, 1995.

by adherents to the guidelines procured by the National Academy of Science. Mann obliged.

As soon as McIntyre and McKitrick began reassembling Mann's data, they noticed a variety of sloppy errors, including mislabeled and obsolete data and unexplained truncations (involving numbers with decimal points). When these errors were corrected and run through the computer, amazingly the sizzling blade of the hockey stick *disappeared.* A report noting their astounding results was published that October.[69]

Nature was informed of the now published glaring errors, and, after their own investigation, ordered a list of corrections from Mann, which were not supplied until years later.[70] Attempting further to uncover Mann's methodology, McIntyre and McKitrick requested Mann's computer codes. Mann refused the request. Acting on an earlier suggestion from Mann that the University of Massachusetts' computer archives housed his original data, they went through the archives for a month, only to discover that they had been led on a wild goose chase—the data they were originally working with was the same information found in the archives. Mann seemed to have purposely provided a bad lead. However, as is often the case in investigative work, the relentless McIntyre and McKitrick got lucky and stumbled upon some buried computer codes in the archives that turned out to be the very ones Mann had used to create his original analysis.

After correcting the striking errors in the data, and using Mann's own data-mining algorithm, the hockey stick model was broken—there was no longer a spike on the right side of the graph, and the centuries of past warming and cooling became evident. When the bad data was re-added, the wild, misleading hockey stick again appeared.

The lobotomizing hockey stick had been snapped like a cheap pencil.

McIntyre and McKitrick contacted *Nature* with their extensive results, requesting the opportunity to show the world their findings. Despite the journal's own admission that bogus information had been presented by Mann in his original paper, after an eight-month reviewing process, *Nature* refused to publish the damaging report. The reason, according to

---

69 McIntyre and McKitrick, "Corrections to the Mann, et al., Proxy Data Base and Northern Hemisphere Average Temperature Series, *Environment and Energy* 14 (6), 1998, pp. 751-771.
70 Mann, Bradley, and Hughs, 2004, Corrigendum, *Nature,* July 1, 2004, p. 105.

McKitrick, was that "[t]hey concluded it could not be explained in the 500-word limit they were prepared to give us, and one of the referees said he found the material was quite technical and unlikely to be of interest to the general readers."[71]

Years later, CRU team members were still so hacked off at McIntyre and McKitrick's tireless efforts to obtain Mann's data sets, that in one October 2009 email to Phil Jones excuses are made for withholding data:

> . . . the issue of with-holding data is still a hot potato, one that affects both you and Keith [and Mann]. Yes, there are reasons—but many *good* scientists appear to be unsympathetic to these. The trouble here is that with-holding data looks like hiding something, and hiding means (in some eyes) that it is bogus science that is being hidden.

Numerous others have debunked Mann's hockey stick model, yet the Stick continues to be touted by global warming advocates as the gospel. Check your kid's science textbooks, it's sure to be there as well—underscoring the fantasy that this is the hottest weather *ever*, because the inconvenient facts are being left *out*.

---

71 McKitrick, Ross, "What is the Hockey Stick Debate About," APEC Study Group, Australia, April 4, 2005.

3

RIGGED RECORD

... and thus in the primitive simplicity of their minds they more readily
fall victims to the big lie than the small lie ... "

—A. Hitler, *Mein Kampf*, Volume 1, Chapter 10

"105° TOMORROW? We'll be sending you out live," my boss informed me.

The San Francisco Bay Area is known for its diverse microclimates. In the summer, temperatures are often near 60° for folks near the ocean, 70–80° for the millions who live in and around the moderating waters of the Bay, and, in the most inland regions, 90° plus; and during even a modest heat wave, inland temperatures easily top the century mark.

"Let me guess, the bank in Walnut Creek?" I said sarcastically. I had been through this drill many times.

"Perfect location," the boss replied, as if the idea was novel. "A lot of viewers with ratings meters out there too."

Walnut Creek is an upscale town 30 miles east of San Francisco. It's sheltered from the cooling influences of the coast and the Bay by the rapidly-rising Diablo mountain range. As a result, in the summer that region can uniquely bake. The bank not only referenced the name of the town, but had a thermometer that was predictably several degrees too high, thanks to the heat-absorbing black asphalt on the adjacent multilane street and the pavement of a nearby parking lot. 105° would easily read 110°. Every time I did this routine I would always try to quickly explain the misleading reading for our audience, but it didn't matter—the visual was everything. Viewers would no doubt comment, "Did you see that? 110 in Walnut Creek!"

Everyone in the newsroom knew extreme weather always sold well, especially with the increasing hype over global warming. And the deceptive bank thermometer? It bothered me, the uptight weather guy, but the newsroom was probably right—most folks casually watching the show would never really discern the difference.

FAKING THE TEMPERATURE

The tragedy is, it's not just faulty bank temperatures that are throwing out bogus readings; the earth's entire temperature record is rigged. The public naïvely assumes that the atmosphere is accurately measured by trustworthy scientists who are concerned about precision and credibility— that is not wholly true. Like Mann and his Hockey Stick model, the handful of "experts" entrusted with monitoring the planet's temperature are agenda-driven apparatchiks who are constantly attempting to tweak the numbers upward.

The actual temperature record of the earth is relatively recent. Galileo invented a rudimentary water-based thermometer in 1593, which, for the first time, allowed temperature variations to be accurately measured. In 1714, Gabriel Fahrenheit devised the first mercury thermometer—the same technology we commonly use to measure temperature today. By the 1800s, mercury thermometers were being mass produced and used in measuring local temperatures throughout the world. However, besides a thermometer record that merely stretches back two centuries, there are other glaring inconsistencies in the historical, global temperature record. For one, thermometer coverage around the world has always been distressingly sparse. Secondly, because of the ravages of war and political upheaval, the logs that do exist back to the nineteenth and early twentieth centuries are filled with glaring holes of missing and unreliable data. The exception of course is the United States—successfully defending our national sovereignty has allowed us to preserve consistent temperature archives.

There are 1,221 official, government-sanctioned weather monitoring stations that have been recognized as a part of the U.S. Historical Climatology Network (USHCN). These stations all possess uniform equipment and guidelines in order that their records remain consistent and historically accurate. Most possess entries that date back to the 1800s. The beauty of this network is that, in so many cases, the environs where the thermometers are housed has changed little over the decades, thus providing critical, accurate data, which is especially useful when determining major long-term trends.

The USHCN sites are individually and manually maintained by their perpetual stewards: farmers, ranchers, firehouse staff, municipal workers, students, and park rangers. There is no automation with this network; each thermometer is visually observed twice daily to accurately determine the maximum and minimum readings (the thermometers possess a clever device that marks the high and low reading each day, hence, eliminating any guesswork). Having met many of these USHCN observers, I can say with confidence that the majority are reliable weather enthusiasts who have a great respect for the important data they collect. Though far from perfect (as you will see), the USHCN is the world's most complete record of modern temperature.

Beyond the Historical Network we also have a system of about 1,000 automated weather sites known as the Automated Surface Observation System (ASOS). These sites are funded and maintained by the federal government in conjunction with the National Weather Service, Federal Aviation Administration, and the Department of Defense. Most of the ASOS monitors were only placed online in the 1980s, with an additional 400 having been added since 1998. Unlike the USHCN sites, which primarily register temperatures in more rural areas, the ASOS data are mostly taken in urban locations, including airports and downtowns—settings that present abnormalities in the record. Like the bank thermometer, ASOS can produce irregularly high temperatures, reflecting what we refer to as the "urban heat island" effect. Also, given the fact that these monitoring stations have only been in operation for a couple decades or less, they are establishing a new, and warmer, history—one that conveniently corresponds with the Hockey Stick graph.

Even though the United States possesses the most thorough network of recording devices, the coverage is scant. In some broad regions of the country, there is only one thermometer representing thousands of square miles; in other sectors, like the highly urbanized northeastern states, there are hundreds of official temperature sites packed together.

In addition to the USHCN and ASOS data collected in the United States, globally, temperatures are also culled from a pitifully skimpy network of ocean-based weather stations, plus a collection of unequally distributed, rag-tag readings scattered around the world—the largest percentage having been installed since the 1980s. Besides inequitable distribution, a conspicuously consistent problem is that overall quality control for this relatively new international mesh is sorely lacking: otherwise-devoted observers like Achmed in Afghanistan is running from the Taliban, Akoko in Africa is dodging warlords, Uri in Ukraine hasn't been paid for months, and the broken thermometer on the buoy in the middle of the Pacific won't be fixed by the U.S. government for two years—oh, and by the way, in the meantime, no one is even aware it's outputting bad data.

There is also a third method of taking the earth's temperature, and it is certainly the most effective method—satellite. Since 1979, the National Aeronautics and Space Administration (NASA) has been effectively using satellites to measure the air temperature directly above the earth's surface (the lower troposphere). These readings have been verified accurate via data collected from weather balloons, which are routinely released into the sky once or twice daily at hundreds of locations around the world. Each balloon carries equipment that radios a signal back to a host of cooperating scientists, who then collect the data and record a variety of weather information. Unlike the ground- and sea-based observation stations, satellites are able to provide 100% coverage of the earth. Therefore, if one really wanted to make pronunciations about global warming over the past 30-plus years, it would make sense to chuck all but the satellite data. However, the wizard in charge of the world's temperature won't go there.

MEET THE WIZARD

The recognized international arbiter of the global temperature record is NASA. With a variety of state-of-the-art satellites, computers, man power, and general oversight of the USHCN, ASOS, and the satellites, NASA's science division consumes nearly a third of the agency's whopping $18 billion annual budget and is deemed the final authority when it comes to backing Al Gore's whopper that "the earth has a fever."[72]

As is the case with all government agencies, maintaining a budget is critical. The bureaucrats at NASA boast of their obvious needs for cash: completion of the International Space Station, furthering the Space Shuttle Program, and, of course, nowadays, preventing the world from spontaneously combusting into a ball of flames. The man responsible for the latter is Dr. James Hansen, for nearly 30 years the director of NASA's Goddard Institute for Space Studies (GISS). Hansen is a zealous promoter of global warming, and, since the 1980s, has been able to keep the funds flowing—both into NASA, as well as into his personal pocket—to study the world's climate.

For the record, Dr. Robert Jastrow, a founder of the Goddard Space Institute, reportedly said to a trusted friend, toward the end of his life, that he didn't have many regrets, but his "one true" regret as a professional researcher was handpicking Jim Hansen to be his successor. Jastrow's friend was Dr. Willie Soon, the aforementioned Harvard-Smithsonian astrophysicist, who shares Jastrow's distain for Hansen's distortion of the scientific process.[73]

A slick marketer, Hansen seems to possess an insatiable appetite for media attention—especially when the person asking questions is favorable to his point of view. In 2007, Hansen agreed to an interview on a rooftop in downtown San Francisco conducted by a counterculture, Internet-based outfit called TUC Radio (TUC is an acronym for "Time of Useful Consciousness"—the time between the onset of oxygen deficiency and the loss of consciousness).[74] During the interview Hansen hardly sounded like an honorable director of a U.S. government agency,

---

72 Gore has said, "the earth has a fever" many times, but in my opinion, the most ridiculous utterance came in his speech before the Nobel Institute, upon receiving his Peace Prize, December 10, 2007.

73 Dr. Willie Soon, "Endangering the Polar Bear," speech presented at the Doctors for Disaster Preparedness annual conference, Mesa, Arizona, 2008.

74 Maria Gilardin, TUC Radio Founder, February 12, 2008, http://www.tucradio.org/about.html.

but more like Marxist community agitator: "I tell young people that they had better start to act up. Because they are the ones that will suffer the most. Many of the changes will take time, but we're setting them in motion now. We're leaving a situation for our children and grandchildren which is not of their making, but they're going to suffer because of it. So I think they should start to act up and put some pressure on their elders, and on legislatures, and begin to get some action."[75]

Prior to the interview, did Dr. Hansen make it clear that all his comments were his own and not representative of NASA? If he didn't, he should have been fired immediately—especially given past concerns about his proclivity for popping off like a nitwit.

Early in 2006, a major story in the *New York Times* pointed a finger at the Bush Administration for supposedly trying to censor Hansen. In part, it read:

> The scientist, James E. Hansen, longtime director of the agency's Goddard Institute for Space Studies, said in an interview that officials at NASA headquarters had ordered the public affairs staff to review his coming lectures, papers, postings on the Goddard website and requests for interviews from journalists.
>
> The top climate scientist at NASA says the Bush administration has tried to stop him from speaking out since he gave a lecture last month calling for prompt reductions in emissions of greenhouse gases linked to global warming.[76]

Can one hardly blame the administration for wanting to review his content? As a NASA director, his role should be collecting data and truthfully sharing results, not trying to influence policy, legislation, and encourage the masses to "act up."

Congressman Darryl Issa (R-San Diego) confronted Hansen on his continual talking out of turn. During a hearing on Capitol Hill regarding

---

75 Interview with Maria Gilardin, TUC Radio, San Francisco, December 2006.
76 Andrew Revkin, "Climate Expert Says NASA Tried to Silence Him," *New York Times*, January 29, 2006.

Hansen's alleged abuse of his government status, Issa said, "You're speaking on federal paid time. Your employer happens to be the American taxpayer."[77]

Issa revealed that an Internet search unveiled more than 1,400 news stories written over the past year in which Hansen was quoted as complaining about being censored by Bush. According to the Associated Press: "Hansen said . . . as a matter of free speech, government scientists should not be restrained in their remarks or have public affairs officers listening in on interviews."[78]

Congressman Issa was correct—government workers like Hansen are being paid by you and me, and should keep their mouths zipped. Bureaucrats should not be allowed to use their job as a soapbox to further their personal agendas.

However, now that Obama is the boss, Hansen obviously is emboldened and is barking like a carnival peanut vendor. In February 2009 Hansen wrote an op-ed in the London *Observer*, recklessly claiming, "The trains carrying coal to power plants are death trains. Coal-fired power plants are factories of death."[79] A month later he had the gall to front a YouTube video[80] produced by Capitol Climate Action, calling for what the group described as "mass civil disobedience at the Capitol Power Plant, Washington, D.C."

And, just like the peanut vendor who is skimming a commission, Hansen's activities and self-promotion has enhanced his personal bottom line. In 2001, the Heinz Foundation "awarded" James Hansen with a payment of $250,000 for helping make "global warming" a household phrase. According to the foundation: "It was Dr. Hansen who, in the sweltering, drought-scorched summer of 1988, went where few scientists were willing to go—before Congress, to explain just how serious the potential for global warming truly was."[81]

---

77 Joseph Hebert, Associated Press, March 19, 2007.
78 Ibid.
79 James Hansen, "Coal-Fired Power Plants are Death Factories," *The Observer*, February 14, 2009.
80 Dr. James Hansen, "A Call to Action on Global Warming," youtube.com.
81 Commemorative Brochure, 7th Annual Heinz Foundation Awards, 2001.

The Heinz Foundation, directed by the wife of U.S. senator and former presidential candidate, John Kerry, is widely known for its support of radical environmental causes. Predictably, Hansen also endorsed John Kerry for president in 2004.[82]

The quarter of a million booty from Heinz was just a tease of additional monies to come. In 2007, Hansen split a $1 million award from the Dan David Prize's category of "Future Quest for Energy" (layman's translation: "a world without CO_2-generating fossil fuels"). In addition Hansen is also said to have acted as a paid consultant to Al Gore during the making of his global-warming film, *An Inconvenient Truth*, and even personally promoted the film during a large NYC showing.[83]

James Hansen has allegedly received hundreds of thousands of additional dollars to further politicize the issue of global warming. According to *Investor's Business Daily*:

> How many people, for instance, know that James Hansen, a man billed as a lonely "NASA whistleblower" standing up to the mighty U.S. government, was really funded by [George] Soros' Open Society Institute (OSI), which gave him "legal and media advice"? That's right, Hansen was packaged for the media by Soros' flagship "philanthropy" by as much as $720,000, most likely under the OSI's "politicization of science" program.[84]

Hansen denied any relationship with the extreme left-leaning George Soros and his Open Society Institute, but *Investor's Business Daily* refused to back off on their story.

With that kind of cash allegedly lining his pockets, why would Hansen ever allow the data he's charged with managing to point to anything but disaster?

---

82 U.S. Senate Committee on Environment and Public Works, Press Release, BROKAW'S OBJECTIVITY COMPROMISED IN GLOBAL WARMING SPECIAL, July 11, 2006.
83 Michael Asher, *Daily Tech*, Blog: Science, Update, "NASA, James Hansen and the Politicization of Science," September 26, 2007.
84 *Investor's Business Daily*, Editorial, September 24, 2007.

RIGGED RECORD

The computers under James Hansen's guard are annually fed the temperature record from the United States Historical Climate Network. His NASA division also inputs the data from ASOS and the global temperature network. Hanson is quite intimate with both of them in that he was consulted extensively during their establishment—it's kind of like owning a major-league baseball team and designing the outfield walls of your new ballpark to best suit your pricey, left-handed home run slugger.

The satellite data is also under Hansen's charge and is grossly misused.

As stated, the world's most accurate historical temperature record is the USHCN. If you were to take all the data from its 1,221 locations and average them, you would see a mere increase in temperature of only one-half degree Fahrenheit since 1900. However, even with these archives, there are obvious problems. Some of the thermometers now carelessly stand next to machinery, heating vents, and port-a-potties. A few lazy stewards were discovered to have placed light bulbs in the thermometer shelters in order to log the afternoon's maximum temperature after sunset . . . forget to flip the switch off and you've ruined the next day's minimum temperature. But a more common problem has been that many of the USHCN stations have seen their once idyllic environs morph into urbanized locales, creating the heat island effect. In a travesty of sound science, these artificially enhanced records are included in determining historical climate trends—making a case for global whiners like Hansen and Gore.

New York City provides us with an excellent example. Readings in Central Park have been regularly measured since 1835, when the city's population had just surpassed 200,000. Today, Central Park is surrounded by a metropolis filled with some of the world's tallest buildings; a massive underground subway system; extensive sewer systems; power generation facilities; and millions of cars, buses, and taxis transporting 8 million residents and 45 million tourists each year. And, as one might expect, the Central Park historical temperature plot illustrates an incredible warming increase of nearly 4°F since the 1800s (see Figure 2.3).

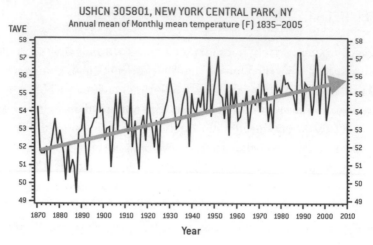

USHCN 305801, NEW YORK CENTRAL PARK, NY
Annual mean of Monthly mean temperature (F) 1835–2005

Figure 2.3 Temperature Readings for Central Park, New York City
Source: CN Williams Jr., MJ Menne, RS Vose, DR Easterling, NOAA, National Climatic Data Center, Asheville NC

Now, as a comparison, let's examine an equally old temperature record collected just 55 miles away from Central Park at West Point Military Academy. Like their New York state counterpart, West Point readings also have been meticulously maintained since 1835, but the environment surrounding the thermometer shelter has experienced minimal manmade interference compared to the one in Central Park (see Figure 2.4).

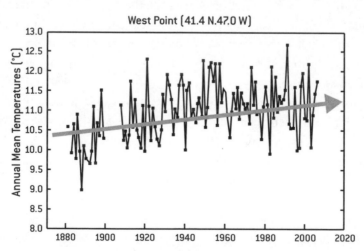

Figure 2.4 Temperature Readings at West Point
Source: CN Williams Jr., MJ Menne, RS Vose, DR Easterling, NOAA, National Climatic Data Center, Asheville NC

In contrast to the Central Park graph, the West Point graph shows a minor increase of only about one-half degree Fahrenheit over the same lengthy period—quite telling given that these journals began just about the time the Little Ice Age was concluding. James Hansen claims data from Central Park and similarly tainted locations are massaged to account for the heat island, before being imputed into his annual calculations, but the rub is, he's the slick masseur pressing where he pleases—a more virtuous plan would be to toss the tweaked readings out.

There are also the additional problems that no mathematical massage can possibly solve. A classic example involves the temperature station charged with monitoring the beautiful Lake Tahoe, California, basin. This is a region so worshipped by environmentalists that even trimming a tree on your own property requires a four-page application and a $53 fee to the Tahoe Regional Planning Association (the tedious bureaucratic process also includes clearing *dead* trees). Local eco-activists constantly grouse about the global warming that has occurred in their backyard—and it's obvious why.

The daily temperature for the basin is taken in Tahoe City, California. The readings are a part of USHCN. When locals look at the record, they are immediately agasp and motivated to whip out a checkbook to fire off a contribution to Al Gore's vacation fund. According to the data, since the early part of the twentieth century, temperatures in the Tahoe Basin have risen close to 4°F (2°C)! (See Figure 2.5.)

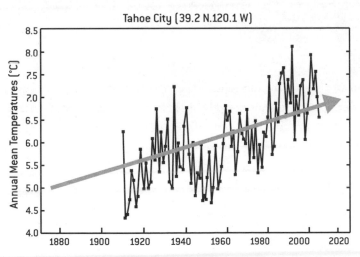

Figure 2.5 Temperature Readings at Lake Tahoe, California

No wonder the locals are in a tizzy. But what they may not know is that the data for the Lake Tahoe area is wholly corrupt.

My friend and colleague Anthony Watts has been doing a yeoman's job in determining how many of the Historical Network's weather sites have been ruined. A TV and radio meteorologist by vocation, he has worked diligently to produce SurfaceStations.org, a premier website dedicated to sleuthing out the bad weather sites in the USHCN (as well as in the ASOS network).

The photo below illustrates why the Tahoe City site is an utter disaster. A 55-gallon drum used for incinerating trash is located five feet from the thermometer shelter. Judging by the rust, the drum appears to have been there for many years. Adjacent to the incinerator is an assortment of heat-absorbing metal junk. In addition, during the Eighties, a tennis court was built some 25 feet away (by regulation, USHCN weather stations are to be 100 feet from any paved surface). The gravel surrounding the chain-link fence is a well-used parking lot that is packed in the summer, as Lake Tahoe is just a short walk away (see Figure 2.6).

Figure 2.6: Weather Station at Tahoe City, California

Source: Photograph courtesy Anthony Watts, SurfaceStations.org.

Also worth noting, adjacent to the right fence post is a compartment that offers plastic doggy poop bags. It is no wonder that this recording station, which provides the "official" temperature for pristine Lake Tahoe, and is included in the global record, *stinks!*[85]

---

85 The 55-gallon drum was finally removed just prior to publication.

THE HOTTEST WEATHER—EVER

Gore and Hansen lead the ignorant to believe that this is the hottest weather ever, blaming it all on the rampant production of carbon dioxide caused by man's insatiable appetite for fossil fuels.

We'll dissect this lie in great detail in the following chapter, but you must realize that CO_2 only accounts for less than 1% of the gases in earth's atmosphere and of that sliver, only 3% is created through anthropogenic means. Additionally, while carbon dioxide has been steadily increasing at a snail's pace since the end of the Little Ice Age, the claim that the gas caused a post-1970 hockey stick spike and the hottest temperatures ever is pure chicanery. And even if it were true, how does that explain the hottest weather ever recorded, back in the 1930s—when CO_2 levels from fossil fuels were immeasurable?

The summer of 1930 marked the beginning of the longest drought of the twentieth century. From June 1 to August 31, 1930, Washington, D.C. experienced 21 days of high temperatures of at least 100°. During that record-shattering heat wave, maximum temperatures were set on nine different days—records that remain unbroken more than three-quarters-of a century later! In 1934, bone dry regions stretched from New York, across the Great Plains, and into the Southwest. A "dust bowl" covered about 50 million acres in the south-central plains during the winter of 1935–1936. In some areas, the drought never broke until 1938.

According to the National Climatic Data Center, 1936 experienced the hottest overall summer on record in the continental United States. In fact, out of 50 states, 22 recorded their all-time high temperature during the 1930s, including:

*110° Millsboro, Delaware, July 21, 1930
*100° Pahala, Hawaii, April 27, 1931[86]
*109° Monticello, Florida, June 29, 1931
*118° Keokuk, Iowa, July 20, 1934
*111° Phoenixsville, Pennsylvania, July 10, 1936

---

86 Yes, I am aware that Hawaii was not granted statehood until August, 1959.

*120° Seymour, Texas, August 12, 1936
*121° Steele, North Dakota, July 6, 1936
*117° Medicine Lake, Montana, July 5, 1937.

One might fabricate the argument that the incredible temperature rise in the 1930s coincided with the exuberant industrialization that was occurring in America; after all those new smokestacks were producing heat-trapping CO_2. But that's faulty logic. During the following three decades, while carbon dioxide levels increased, temperatures *decreased.*

POST TOASTY THIRTIES: COOLING, WARMING, AND COOLING AGAIN

According to NASA, the average temperature on the planet between 1940 and 1970 dropped .18°F (.1°C). Though it doesn't sound significant, the decrease was enough for major media outlets to jump on the ice age bandwagon and make pronunciations like those at *Time* magazine who exclaimed: "Climatological Cassandras are becoming increasingly apprehensive, for the weather aberrations they are studying may be the harbinger of another ice age."[87]

This 30-year cooling trend has always been bothersome to most activist scientists because while temperatures were slightly dropping, carbon dioxide was slightly increasing. If CO_2 really is a leading indicator of temperature change, the earth should not have been swamped in a multidecadal cool down. In a leaked CRU email, written to Phil Jones in September 2009, we read of a bald-faced attempt to manipulate data from this era:

So, if we could reduce the ocean blip by, say, 0.15 degC, then this would be significant for the global mean—but we'd still have to explain the land blip. I've chosen 0.15 here deliberately . . . It would be good to remove at least part of the 1940s blip, but we are still left with "why the blip."

By reversing, or moderating, the cooling "blips" of the 1940s, these

---

87 *Time,* June 24, 1974.

charlatans could further lobotomize the record, making the following 1970–1998 minor rise in temperature, which totaled .34°F (.19°C), appear more significant.

However, since 1998, the temperature trend—even by NASA's often biased standards—has been relatively flat, with a notable downward orientation commencing in 2007. In fact, when the corrupted data from the Historical Network is excluded and only the most trusted sites observed, since the onset of a noticeable increase in CO_2 in 1930 there has actually been a net *decrease* in temperature in all quarters of the continental United States.

Yet, despite the facts, temperature keeper James Hansen remains on the stump, blubbering to the press: "We have only four years left...."[88]

Four years to do what—continue to play a leading role in duping the public so he, Gore and a few others can rake in the lucre?

And look at how effectively this mad maestro plays the media. Just a few days into 2007, the *Washington Post* ran a front page story that read, "Last year was the warmest in the continental United States in the past 112 years—capping a nine-year warming streak 'unprecedented in the historical record' that was driven in part by the burning of fossil fuels, the government reported yesterday."[89]

So, according to the *Post*, 2006 was the hottest in 112 years. A couple days before their story ran, Reuter's news service ran this headline: "2007 is expected to be the hottest year on record." The report stated, "This year is expected to be the hottest on record worldwide because of global warming...."

Talk about a prediction! Only four days into the 2007 year and a bet-the-farm prophesy had been issued insisting that the next 361 days would add up to the hottest ever in the history of the planet—this following the supposed record-breaking 2006 scorcher. Back to back record claims . . . and the mainstream media ran both *without challenge*.

How many of those same news outlets were honorable enough to present the truth about 2006 and, eventually, 2007?

---

88 Robin McKie, "We Have Only Four Years Left," *The Observer*, Sunday, January 18, 2009.
89 Mark Kaufman, "Climate Experts Worry as 2006 Is Hottest Year on Record in US," *Washington Post*, Wednesday, January 10, 2007, A01.

VIRTUALLY NONE

Re-enter Steve McIntyre, who contacted NASA to inform them that their 2006 calculations were in error. The year of 2006 may have been a hot one, but it was certainly not the front runner that everyone proclaimed. McIntyre's calculations forced Hansen and company to quietly republish the record to officially rank the 20 hottest years as follows:

20 HOTTEST YEARS[90]

1934	1939
1998	1987
1921	2001
2006	2007
1931	1986
1999	2005
1953	1946
1990	1991
1938	1933
1954	1981

Interestingly, so many of the global warming hawkers get the warmest years wrong. In both his film and on the road, Al Gore has claimed, "If you look at the ten hottest years ever measured, they've all occurred in the last 14 years. And the hottest of all was 2005."[91]

Does 2005 even appear in the top ten? No. 2005 is actually number 16. I honestly don't know where Gore obtains his data.

And 2007? It should actually be remembered as the Year of the Big Chill. In Buenos Aires, the first snow fell since 1918. In Peru, crops failed, livestock perished, and the Peruvian government declared a state of emergency as hundreds died in their worst cold snap ever on record.

---

90 Goddard Institute for Space Studies, updated from their database May 27, 2009, http://data.giss.nasa.gov/gistemp/graphs/Fig.D.txt.
91 Interview with Al Gore, *Larry King Live*, CNN, June 13, 2006.

Townsville, Australia experienced the longest period of continuously cold weather since 1941. Snow fell for first time in a century in Baghdad, Iraq. A record cold gripped China. Sea ice in the Arctic grew at a record pace. Snow fell in parts of South Africa for the first time in 30 to 40 years. By the end of 2007, the whiners were icily silent—or at least, they should have been: 2007 didn't even make it into the top 10 hottest years; in fact a period of cooling was clearly underway.

Scrambling like a witch doctor searching for another pin to stick in his global warming voodoo doll, Hansen frantically came out with a press release from his GISS office, on November 10, 2008, claiming the previous month was the hottest October in history. This was a shock for many given that in October the Swiss lowlands received the most snow for that month since 1931, areas of Florida broke 150-year record minimum temperatures, and in an ironic event, London experienced its first October snow since 1922—the same day the House of Commons was debating Global Warming.

Enter my friends Anthony Watts and Steven McIntyre who busted Hansen for "indecent exposure." They performed their own detailed analysis of Hansen's reported data and discovered disturbing details: GISS temperatures taken from across a broad swath of the earth in Russia were not readings from October—they were a repeat of the data culled from that same location in September!

GISS was forced to fess up. They quietly retracted the figures and expunged the November press release from their website.

2008 ended up as one for the records, going down as the coldest of the new millennium, with 2009 even colder still. Yet, Hansen's insane crystal gazing continues to be gobbled up by global whining fools, including his famous million-year forecast:

> The average temperature is now 0.8 degrees Celsius higher than in the last century with three-quarters of the increase happening in the last 30 years. But . . . there's another half-degree Celsius in the pipeline due to gases already in the atmosphere, and there's at least one more half-degree to come due to power plants which we're not going to stop immediately. Even if we decide now, we've got to slow down as fast as is practical, there's

still going to be enough emissions to take us to the warmest level that the planet has seen in a million years.[92]

Temperature *in the pipeline . . . warmest level that the planet has seen in a million years?* This isn't the speech of a scientist—this is the mesmerizing prattle one would expect from a pay-for-play propagandist.

Most TV weathercasters provide you a five-day forecast to cap their evening presentation. Some will offer a seven-day forecast, and a few will stretch it out to ten days. While these extended forecasts are always a big deal to the behind-the-scenes producers, they are usually deemed unprincipled by the smiling weather person because the predictions are often so inaccurate. When I was doing TV weather, I always felt pretty confident with my two-day forecast, but honestly, I would be hesitant to bank on my five-day prediction, let alone bet the family farm on my seven- to ten-day forecast. Why? Because the computer models are imperfect. Yet, the forecasting models used by the TV weatherperson are essentially the same models being run by Hansen's colleagues at NASA. If a television weather dude won't bet on his own five-day forecast, why should our government listen to Hansen and create policy based on his million-year guess?

SUMMARY OF GLOBAL TEMPERATURE

Here is a key summary of the global temperature record:

- Commencing with the end of the Ice Age to approximately 900 A.D., the earth stumbled out of its deep freeze.
- At the peak of the Medieval Warm Period, 900–1300 A.D., best estimates suggest the temperature rose to slightly more than 2°F (1.2°C) warmer than today.
- During the Little Ice Age, 1350–1800 A.D., the warmth of the Medieval Warm Period was wiped out, with temperatures

---

92 Interview with James Hansen, "We Need to Take Action Soon," *Der Spiegel*, April 10, 2007.

at one point falling to about 2°F (1.2°C) cooler than today.

- Following the Little Ice Age, temperatures stabilized for approximately 50 years.
- Between 1850 and 1940, temperatures increased about 1°F (.6°C).
- From 1940 to 1970, the earth cooled .18°F (.1°C).
- A minor warming of a mere .34°F (.19°C) occurred between 1970 and 1998, as verified by NASA (many activists both inside and outside NASA have tried to skew the NASA data—which includes satellite records—higher, but thanks to the work of NASA-affiliated scientists Dr. Roy Spencer and Dr. John Christy at the University of Alabama's Earth System Science Center, those activist efforts proved futile).
- Average global temperatures have not warmed since 1998.
- As of 2007, a compilation of all temperature records indicates a warming since the mid-nineteenth century of slightly more than 1°F (.7°C, though James Hansen contends the figure should read .8°C).

However:

- According to the USHCN archives, the temperature has warmed only .5°F since the mid-1800s.
- Tossing out corrupted USHCN data, there has actually been a net-*cooling* since 1930—during the same period in which atmospheric CO_2 has noticeably increased.
- Since 2007, global temperatures are engaged in a significant downward spiral, with government data illustrating a bit more than a 1°F (.65°C) *drop* in temperature between 2007 and 2008 alone.[93]

Like Hansen, Al Gore plays a phony game of climate clairvoyant as well. Dusting off his crystal ball in *An Inconvenient Truth*, Gore claims that because of CO_2's impact on Earth's atmosphere, sea levels will rise

---

93 Hadley Climate Research Unit Temperature (HadCRUT) Anomaly plot, 1988–2008.

by as much as 20 feet, Arctic and Antarctic ice will likely melt, heat waves will become "more frequent and more intense," and "deaths from global warming will double in just 25 years—to 300,000 people a year."

And with what honest scientific facts and measurable observations does he draw his conclusions? None.

All *poof*, but no proof!

4

DESIGNER POLLUTANT

People always have been the foolish victims of deception and self-deception in politics, and they always will be. . . .

V. I. Lenin, *The Three Sources and
Three Component Parts of Marxism* (1913)

IN 2007, Al Gore rolled into Washington, D.C., to lecture key congressional committees about the perils of climate change. Obviously Gore was no stranger to this venue. He served in both wings of Congress for a total of 16 years before becoming Bill Clinton's vice president for eight. However, this was the first time he had spoken on Capitol Hill since receiving his Hollywood Oscar, his Swedish Nobel, and the under-reported millions of American greenbacks from a variety of sweetheart financial deals.

The media attention for his testimony was akin to the arrival of a universal superstar, and when he finally spoke, his challenge was ecclesiastical: "Our world faces a true planetary emergency. I know the phrase sounds shrill, and I know it's a challenge to the moral imagination." The problem, Gore said, was "carbon dioxide." Then Gore dropped this chilling and sobering line: "There are times when a small group has to make difficult decisions that will affect the future of everybody."[94]

---

94 Al Gore, speaking before a joint session of the United States House of Representatives Committees on Energy and Commerce, and Science and Technology, March 21, 2007.

The difficult decision Gore recommended was imposing upon the populace an immediate cap on the nation's carbon dioxide emissions and commanding further reductions of 90% or more (from today's levels) by 2050. Such legislation should cause one's "moral imagination" to envision life in a third-world gulag.

Of course, none of the media lackeys covering this event used their own imagination to put Gore's solution into relatable terms, but an honest evaluation would reveal that given the population is always increasing in the United States (according to the Census Bureau, we experience a net gain of one person every 13 seconds, which equals 6,646 new people in this country daily and nearly 2.5 million annually), an immediate cap on CO_2 would demand an immediate *reduction* in lighting, heating, and cooling your home; your driving would have to significantly decrease *today*; and industry would have to radically scale back—*now*.

Further, Gore's machinations for 2050 are untenable. Census data projects that by 2050, the population of the United States will have increased from 310 million to 404 million. Presently 34% of CO_2 emissions are derived from transportation, 40% from electricity generation, and 25% from business and industry (including natural gas to heat homes). Add meeting the lifestyle needs of 94 million new people into the mix, and a 90% reduction in carbon dioxide emissions by 2050 essentially requires the elimination of *all* fossil fuels.

So, if fossil fuels become *verboten*, from where will our future energy derive? Certainly not from alternative sources such as nuclear, wind, and solar. As you'll later discover, most so-called renewable energy plans are a pipedream.

Sadly, Congress has listened to Gore, and now his plan has become the federal government's blueprint for projected CO_2 reductions. If successful, by 2050 their vision for America will look like North Korea as presently seen from satellite on any night, with the entire country hidden in darkness, sans an illuminated dot representing their capital city (Figure 4.1).

Gore's lamentations propose not just an impossible proposition, but an immoral imposition . . . because carbon dioxide is *not* a pollutant; it's a vital fertilizer essential to life. In fact, it's no more deadly than water and oxygen.

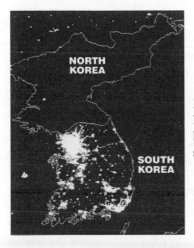

Figure 4.1: Satellite image of an entirely dark North Korea, except for Pyongyang, and a highly illuminated South Korea, below, and a sparsely lit China

LIFE-GIVING CARBON DIOXIDE

Like water and oxygen, there is a finite amount and yet an infinite supply of carbon dioxide. In other words, we are not losing any to outer space, nor is any being added from deliveries by extraterrestrials. The CO_2 that is here, *is here*—and it's not going away—it will just be *stored* in different places.

Observe the wonderful symbiotic relationship among humans, animals, and plants. Humans, and the various species of animals, breathe *in* the oxygen produced by plants. A wondrous transformation then materializes in our lungs, causing us to exhale CO_2 which the plants then "breathe." Just as fascinating is the relationship plants have with the planet's various other sources of CO_2. Decomposing vegetation, the carcasses of dead animals, forest fires, smoldering peat bogs, volcanoes, plowed soil, weathering rocks, human utilization of fossil fuels, and even termites and crustacean shells—all exude carbon dioxide beneficial to the plant kingdom. And the more carbon dioxide in the atmosphere, the more content the plants become—just ask anyone who has worked in a greenhouse. In fact, that is a portion of the carbon dioxide debate no one bothers to address—the plant kingdom would abound if carbon dioxide levels were to increase in the global atmosphere.

Research gathered by Michigan State University professor emeritus of horticulture, Sylvan H. Wittwer,[95] indicates that with a tripling of CO_2, roses, carnations, and chrysanthemums experience earlier maturity, have longer stems and larger, longer-lasting, more colorful flowers with yields increasing up to 15%. Rice, wheat, barley, oats, and rye perform yield increases ranging to 64%. Potatoes and sweet potatoes produce as much as 75% more. Legumes, including peas, beans, and soybeans, show increases to 46%. The effects of carbon dioxide on trees, which cover one-third of Earth's land mass, may be even more dramatic. According to Michigan State's forestry department, trees have been raised to maturity in months instead of years when the seedlings were raised in a tripled CO_2 environment.[96]

So, how could an element so essential to life be vilified as evil? Reckless comments by blathering fools like Gore, or, especially, environmentalist Robert Kennedy Jr.—who, in an interview for *Vanity Fair*'s Green Edition in 2008, likened ending carbon emissions to abolishing slavery—are made to produce fear, guilt, and eventual support from the masses, so these slick Marxist hucksters can play the people like a cheap accordion.

SO LITTLE CO_2

It's astounding to note that, of the gases in our atmosphere, the amount of carbon dioxide is almost imperceptible. By percentage, the gases are ordered as follows:

Nitrogen	78.1%
Oxygen	20.9%
Water vapor	.40%[97]
Argon	.9%

---

95 Sylan H. Wittwer, "Rising Carbon Dioxide is Great for Plants," *Policy Review*, Fall, 1992.
96 Ibid.
97 The total atmospheric amount of water vapor always varies. It is about .40% in the total atmosphere, and 1-4% near the earth's surface.

Carbon Dioxide	.038%
Neon	.002%
Helium	.0005%
Methane	.0002%
Krypton	.0001%
Hydrogen	.00005%
Nitrous Oxide	.00003%
Ozone	.000004%
Carbon Monoxide	just a trace

Carbon dioxide only accounts for a scant 38 thousandths of a percent of our planet's atmosphere.[98] It is known as a variable gas, because, like water vapor, it has historically fluctuated. And what percentage of the miniscule amount of CO_2 is produced by the activities of man, including the utilization of fossil fuels? According to a thorough analysis by the Carbon Dioxide Information Analysis Center, a research wing of the U.S. Department of Energy, it is only 3.207%.[99] All of this hoopla over an atmospheric component so minute, it is difficult to comprehend.

Allow me to repeat this critical fact:

Carbon dioxide comprises 38/1000ths of the earth's atmosphere, and of that amount, a mere 3% is generated by mankind.

---

98 Each time I publically enunciate the decimal .038 I am hit with a barrage of critics who contend it's "38 hundreths." For those who have become a bit fuzzy on their elementary school math, watch an excellent video on YouTube by a teacher known as Roadshow: http://www.youtube.com/watch?v=iaSFUPRReow.

99 "Current Greenhouse Gas Concentrations," Carbon Dioxide Information Analysis Center, U.S. Department of Energy, Oakridge, Tennessee, updated October 2000. I originally retrieved this information in early 2009 when the CDIAC was including "Manmade additions" into their table summarizing the concentration of carbon dioxide in the atmosphere. At the time, the CDIAC stated the total concentration of greenhouse gas was 370,484 ppb, with anthropogenic additions of carbon dioxide totaling 11,880 ppb (3.207%). The CDIAC updated the "Current Greenhouse Gas Concentrations" figures in December, 2009 and no longer include a breakout for "Manmade additions."

And how much has CO_2 increased in the atmosphere over the past 150 years? Approximately 35%. That's all. Slightly more than a one-third increase in a century and a half! And the increase is clearly within the earth's historical norms.

In a must-read eco-thriller, *State of Fear*, the late author Michael Crichton creates a brilliant visual to assist us in wrapping our minds around the components of Earth's atmosphere. On page 387, he likens the atmosphere to a football field. The goal line to the 78 yard-line contains nothing but nitrogen. Oxygen fills the next 21 yards, stretching to the 99 yard-line. The final yard, except for about two and a half inches, is argon, a wonderfully mysterious inert gas useful for putting out electronic fires. About half of the remaining inches are crammed with a variety of minor, but essential, gases. And the last 1.37 inches? Carbon dioxide. The equivalent of 1 inch out of a 100-yard field! If you were in the stands looking down on the action, you would need binoculars to see the width of that line. And the most important point—how much of that last inch is contributed by human activities? The equivalent of a line as thin as a dime standing on edge.

Are you still worried about the dangers of CO_2?

Me, neither—that's why I've written this book.

Are you beginning to understand global warming is a manufactured crisis, the likes of which might actually cause the ghost of Marx to salute its success?

And it only becomes more devious.

GREENHOUSE GAS GAME

The "greenhouse effect" is aptly named. First coined by scientists in the 1800s, it describes the way crucial gases in our atmosphere absorb heat from the sun, thus maintaining an environment appropriate for human habitation. Scientists back then used the term in a favorable way, creating a term that conjured relatable imagery of the warmth experienced in a flower or vegetable greenhouse.

A greenhouse, with its glass walls and roof, allows the sun's heating rays to shine through and enter the otherwise sheltered environment, warming it nicely in comparison to conditions outside the glass.

In addition, the greenhouse traps an entire day's worth of warmth, preventing the heated air from completely radiating back into outer space at night. When soil is brought into the greenhouse, seeds are sown, and irrigation is applied, other factors begin to dramatically warm the artificial environment. The water from irrigation begins to evaporate, creating vapor and increasing the humidity to further warm the surroundings.

Depending on where you reside, you may be well acquainted with the effects of humidity. On a humid summer day you can't move about outdoors without beads of perspiration breaking on your brow. At night, the water vapor-laden air seems heavy, and the temperature has a difficult time dropping to comfortable sleeping levels. This is because humid air tends to retain its temperature. You are experiencing the greenhouse effect working at the micro-climate level—in this case your region of the country.

On the macro-climate level, without the greenhouse effect, the earth would be a ball of ice, void of life—and yet this life-dependant atmospheric factor has become the environmentalists' paramount enemy.

Curiously, research I culled from the Department of Energy fails to list water vapor as a greenhouse gas. This is incredibly disingenuous, given that, in reality, water vapor is the giant of all greenhouse gases, accounting for 95% of their sum. Perhaps water vapor is not mentioned because the computer models—the imperfect utilities that assist both the sincere TV weather guy's five-day forecast, and Jim Hansen's million-year prediction—do *not* consider water vapor in their climate calculations.

Following water vapor, the remaining 5% of the greenhouse gases are, in order, CO_2, methane, nitrous oxide, ozone, and carbon monoxide. However—stay with me here—it must be noted that methane is 21 times more potent than CO_2 when it comes to the greenhouse effect, and nitrous oxide is 310 times more capable of retaining the sun's heat than CO_2. Carbon dioxide is actually a puny player in the greenhouse game.

Since the Department of Energy and other government climatologists choose to ignore the mighty greenhouse effect of water vapor, for the sake of the following proposition, let's eliminate it from the equation and focus only on the remaining 5% of greenhouse gases.

Human contribution of carbon dioxide in our atmosphere is realized primarily through the burning of fossil fuels, but also through important processes like manufacturing cement (the chemical reaction necessary for its production releases CO_2) and even farming (plowing a field exposes microscopic organic matter in the soil, causing the carbon-laden organisms to die, thus releasing CO_2 into the atmosphere). Accounting for the individual concentrations and potencies of the other greenhouse gases (again, sans water vapor), the contribution of CO_2 emissions created by human activity accounts for only *2.32%* of the earth's greenhouse effect.[100]

When we consider water vapor into the math, the anthropogenic carbon dioxide footprint is reduced to a mere *.116%* of the greenhouse effect. Without a calculator, our brains aren't designed to think that small.

Allow me to summarize, because I know you're going to dog-ear this page:

1. CO_2 comprises .038 percent of the earth's atmosphere, and of that amount, a mere 3 percent is generated by mankind.
2. CO_2 emissions created by human activity account for .116% of the greenhouse effect.
3. Since the end of the Little Ice Age (150 years ago) the amount of carbon dioxide has increased 35%, well within historical norms.

The world is being hoodwinked. There is no planetary emergency caused by an abundance of carbon dioxide. And, even if anthropogenic CO_2 was a life-threatening issue, the earth has efficient mechanisms in place to accommodate it.

RECYCLING CO$_2$

The winter snow falls upon the mountain tops, melts in the spring, trickles into a stream, meanders to a river, and eventually fills a larger body of water such as a lake, bay, or ocean. Along the way, some of this

---

100 Again, my calculations are based on 2000 data from the CDIAC. When the CDIAC adjusted greenhouse gas concentrations for the retention characteristics found in methane, nitrous oxide, CFCs, and other trace gases, the total was 509,056 ppb. Thus, adjusted for potency, anthropogenic CO2 (11,800 ppb) would equal 2.32% of all greenhouse gases (ignoring water vapor).

water sinks deeply into underground aquifers, only to be released, in some instances decades later, via a spring or geyser. In addition, each day a fraction of the water atop the earth's surface is warmed by the sun's rays, evaporates, and is caught up in the sky.

Once airborne, this water vapor mixes with the general atmosphere and eventually cools. The atmospheric cooling forces the water vapor to condense into droplets, which gather together, eventually forming a graceful cloud. The cloud conjoins with others, grows, and eventually reaches critical mass, creating a destabilization, followed by precipitation, thus returning the water back to earth in liquid drops, flakes of snow, ice pellets, or hail. This is the earth's water cycle. Life on the planet cannot exist without it. There will never be more or less water in the system. Because of this cycle, water is a resource that is finite yet infinite—it's a zero-sum game. CO_2 works the same way, with one exception; the CO_2 circuit requires more time to complete.

When a major volcano blows its lid on the Pacific Rim, a lightning-induced forest fire rages in the Rockies, or an Indonesian peat bog eternally smolders, huge amounts of long stored CO_2 are naturally released into the atmosphere. The carbon dioxide banked in weathering rocks, decaying coral, and decomposing plants is also constantly meandering through the cycle. Same thing with the carbon cached in fossil fuels; upon consumption it's released into the atmosphere, where it's temporarily held and finally absorbed by a variety of repositories or "sinks."

About these sinks: the most obvious is the atmosphere itself, where, best we can tell, the level of carbon dioxide generally fluctuates between .03–.04%. The largest includes all bodies of water (71% of the earth's surface), which contain vast stores of dissolved CO_2. Dissolved carbon dioxide is used by snails, shellfish, and coral for the formation of their wonderfully hard exoskeletons. The ocean's floor is also rich in sedimentary limestone—a "petrified" modification of CO_2, also known as calcium carbonate (the same stuff you take to ease a tummy ache or heartburn). A major sink often overlooked by the layperson includes the CO_2 that has been petrified over time in the earth's rocks. And a more obvious sink includes the organic carbon compounds found in all things both alive and dead. The dead organic matter includes ancient peat moss deposits, coal seams, natural gas and petroleum reserves, as well as newly fallen autumn leaves, recently felled trees, and even animal and human corpses. Through the process of decay, the carbon stored in all organic substances is released back into the air as inorganic carbon dioxide, to be reworked into the carbon cycle.

There is also a mystery at play in the carbon cycle; a mystery that Marxist scientists loathe (because they prefer to feign supreme knowledge). When an agitation occurs within the carbon cycle—for example, a major volcanic eruption—natural mechanisms seem to maintain the cycle's equilibrium. This was noted when Mount Pinatubo erupted in 1991. I recall showing my television audience the enormous ash plume as witnessed from satellites during that major volcanic blast. A cloud of particulates eventually covered much of the globe. Over the next two years earth's average temperature dropped by one-half degree Fahrenheit.[101] It's theorized that the particulates injected into the atmosphere by Pinatubo were able to partially block the sun's radiation and thus, decrease the global temperature. The mysterious conundrum for scientists is that during the same two-year period, amounts of atmospheric carbon dioxide also *decreased* globally.[102] Could it be that the massive amounts of CO_2 spewed into the sky by one of the most powerful volcanoes in our lifetime was being offset by some natural mechanism like the ocean (the most probable theory)? Advocates of human-caused global warming are quick to emphatically say "no"—because even a "maybe" would blow their anthropogenic theory.

HYPOTHESIS OF OUR LIFETIME

Let's revert to our middle-school science fair analogy: enter an inquisitive youngster, raised by hippies, named Sequoia. Young Sequoia desires to present the following hypothesis: human consumption of fossil fuels

101 "The 1991 Mt. Pinatubo Eruption Provides a Natural Test for the Influence of Arctic Circulation on Climate," Goddard Space Flight Center News Release, March 12, 2003, http://www.nasa.gov/centers/goddard/news/topstory/2003/0306aopin.html.
102 "Large Volcanic Eruptions Help Plants Absorb More Carbon Dioxide from the Atmosphere," Goddard Space Flight Center News Release, December 10, 2001, http://www.gsfc.nasa.gov/topstory/20011210co2absorb.html.

emits CO_2 into the atmosphere, which acts as a greenhouse gas and unnaturally warms the earth's climate.

To go all out in testing her hypothesis, Sequoia would first have to overcome a glaring problem, namely, how would she control the experiment? After all, no one can create an identical earth, which is exactly what would be required to properly run this test. In fact, ideally this experiment would require several earths. The first could be called Earth Gore and would be void of humans using fossil fuels. Over time Sequoia could determine how much the temperature changes now in an oil-, coal-, and natural gas-free world.

Next our student would need an Earth Ehrlich, which would have no living, breathing humans, because humans exhale about 2.8 lbs. each of CO_2 every day, which adds up to over a half-ton per person per year, multiplied by a population approaching seven billion. Plus, the essential activities *Homo sapiens* participate in for survival, like farming food and harvesting trees for lumber and firewood, all include carbon dioxide emissions. Sequoia could really get a beam on how the climate would respond in a people-free environment

However, my educated guess is, even after eliminating the human population, Sequoia would see that CO_2 periodically increases naturally. Paleoclimate researchers are quick to show data illustrating that in the Eocene period (50 million years ago) CO_2 was likely up to six times higher than today. In the Cretaceous period (90 million years ago), it was perhaps as much as seven times higher than today, and in the aptly named Carboniferous period (340 million years ago) carbon dioxide was thought to be nearly 12 times higher than current levels.[103]

Many theorize that dinosaurs were able to grow to such monstrous sizes because the indescribable abundance of CO_2-fed foliage and overall atmospheric conditions present during that era. Certainly the SUV could not be blamed for *those* high levels of CO_2. Dinosaur flatulence, perhaps? We'll never know.

A second major problem sure to be noticed with Sequoia's experiment is that her science fair project is only focusing on carbon dioxide. Inquiring

---

103 C.J. Yapp, and H. Poths, "Ancient Atmospheric CO_2 Pressures Inferred from Natural Goethites," *Nature*, Vol. 355, No. 23, January 1992, pp. 342-344.

minds will certainly want to know the impact of other greenhouse gases on global warming. Research indicates that termites produce more greenhouse gas than human activity ever has (there are 1,000 pounds of termites for each human, and, as mentioned, the punchy methane gas they expel from their little tushies is 21 times more potent than the greenhouse effect from CO_2). How much of the current warming is coming from the termite population or, for that matter, from cow burps and elephant fumes?

Also, water vapor *must* be taken into account. If H_2O makes up 95% of all greenhouse gases, then how have periodic changes in weather patterns, cloud cover, and humidity altered the greenhouse effect in recent decades?

Third problem: temperature data. In order to prove the earth is warming . . . *it actually has to be warming.* As discussed in the last chapter, since 1998, it has not. Using the best data from the U.S. Historical Climate Network, temperatures have actually experienced a net cooling over the past 75 years. And, if the temperature record isn't enough to poke holes in Sequoia's hypothesis, there's another piece of data that certainly will: CO_2 levels, subtly increasing since the 1800s, with a pronounced tick upward beginning in 1930, do *not* coincide with the temperature record . . . despite what she might have seen in her mandatory classroom viewing of *An Inconvenient Truth*.

AL'S MAGIC GRAPH

In all of his presentations, including his movie, Gore simultaneously shows two graphs: one tracking CO_2 over the last 650,000 years, alongside another tracing temperature over the same period, embellished with the aforementioned signature hockey stick spike in recent decades. The casual observer is struck by the way these two graphs seem to mimic one another. I am sure more than a few people have seen the comparisons and concluded, "Dude, CO_2 and temperatures are parallel. I'm purchasing a Prius."

Truth is, in the film Gore employs a slimy psychological tactic we might refer to as "pimping the audience." With a gleam, he addresses the camera and says regarding the graphs, "Incidentally, this is the first time anybody outside of a small group of scientists have seen this image."

Gore only allows a brief glance at his magic graph, making it impossible to dissect 650,000 years of data. However, were the viewer allowed a closer examination, it would reveal that atmospheric concentrations of CO_2 *always* fluctuate.

Carbon dioxide is measured in parts per million (ppm), and currently is at 380 ppm. CO_2 reached a low of about 180 ppm during several periods of extreme cold over the past 400,000 years, but always rose to over 300 ppm between these ice age periods. In fact, according to ice core records[104] at the end of each of the last three major Ice Ages, atmospheric temperatures rose 400 to 1000 years *before* CO_2 levels increased. And, in more recent times, Al's graph reveals that when the temperatures began to warm following the Little Ice Age, an increase in carbon dioxide followed. Likewise, from 1940 to 1970—the temperature dropped while the carbon dioxide levels went up. A similar conflict occurs after 1998.

What Gore's gag graph really illustrates is that the line representing CO_2 is not *leading* the line marking temperature. Instead, the CO_2 line *follows* the increases in temperature. Global whiners aren't observing the details and instead are following sweeping overall trends. From their dogmatic perspective, "temperature has gone up and so has CO_2—there's obviously a connection, so shut up."

But with this type of unscientific cause-and-effect thinking, you could "prove" a lot of things. As CO_2 has steadily risen, so has the incidence of cancer. In fact, during the time at which carbon dioxide has trended upward, child abuse, violent crime, and human trafficking have all increased. Misusing cause-and-effect is an irresponsible way to perform science, but that's exactly what's happening among scientists trying to *prove* global warming today.

MODELS OF PROPAGANDA

Gore, Hansen, Mann, Jones, and the other climate change hawkers continually refer to computer model projections that point to certain calamity unless anthropogenic carbon dioxide is curtailed. However, to

104 Fischer et al., 1999.

hang our collective hats on such a flimsy rack is ridiculous. As we saw with Michael Mann, by inputting selective, biased information, you can demonstrate anything. Similarly, the models that predict weather and climate, both for tomorrow morning's commute and for many years to come, are easily manipulated and rely on grossly insufficient data. Known as the General Circulation Models (GCMs), they are the best tools for prognostication available, but they are attempting the impossible: predicting the actions within a massive "open system."

Imagine, for example, a much smaller open system like the stock market. If I informed you that some brilliant minds (and, no doubt, many of the climate modelers are extremely bright people) had done their very best to create a model to predict how all of the various sectors of the New York Stock Exchange would perform both tomorrow and years from now, would you bet what's left of your 401K on that model's stock picks? Only if you had surplus cash that you didn't mind losing.

In attempting to take on an open system, the GCMs operate on severely limited observed data. As discussed in the previous chapter, the current temperature record of the earth is extremely sparse, unless one relies totally on satellite data, which is quite thorough—but surprisingly, the GCMs do *not*. Artificially warm readings due to the urban heat island effect are manually manipulated by scientists trying to present the models with "the real" temperature. Weather data from a limited set of ships at sea and buoys bobbing on the waves are also inputted into the GCMs—hardly a fair representation of how the oceans are involved in atmospheric interactions.

Moreover, the favored models are shockingly incapable of factoring the effects of water vapor—the granddaddy of greenhouse gases. An even bigger hole in the GCM donut is the sun. Fluctuations in solar radiation are completely left out of the equations. Nevertheless, the mind-numbingly incomplete data stuffed into the models is extrapolated to provide predictions about what is going to happen a day from now, a week from now, and in the case of global warming, *decades, centuries, and millennia* from now.

Another reason to question those who arrogantly point to the current models as barometers of climate catastrophe is that scientists don't know everything—including answers to some of the most basic climate questions.

Many years ago I had the wonderful experience of striking up a friendship with Dr. Sid Serebreny, a pioneering figure in the field of meteorology. He was a popular professor at San Jose State University, a consultant at the Stanford Research Institute (his work there was classified, but he did tell me it had to do with developing satellites that could "see" through clouds), and former chief of Pan Am's flight forecasting center. "Doctor Sid" was also one of the first to discover the jet stream, which he once described to me in a television interview.

During World War II, while Doctor Sid was forecasting for U.S. Army Air Corps pilots flying missions across the Pacific, there was a horrific problem.

"We kept losing planes over the Pacific," he explained. "Aircraft would prematurely run out of fuel because of the tremendous headwinds they were experiencing while flying at 20,000 to 30,000 feet."

The pilots were reporting wind speeds in excess of 200 miles per hour. Dr. Serebreny said headwinds of such velocity were previously unheard of.

"Initially," Serebreny reflected, "many thought the pilots were experiencing equipment malfunction."

Serebreny and colleagues had previously hypothesized that somewhere in the upper atmosphere there had to be a mechanism to drive the earth's colossal weather patterns. Those fierce headwinds over the Pacific led them to significant evidence of their theory—but only after their inadequate weather forecasts resulted in the loss of life, aircraft, and failed missions. Later experiments and research led to conclusive proof of what we now know as the jet stream. However, Dr. Serebreny told me that to this day no honest scientist can definitively conclude whether it is those invisible rivers of wind blowing up to 300 miles per hour that are driving earth's weather machine or if it's the stirring of the mighty ocean currents below, or some combination of both.

However, it's amazing to think that even though the climate of the earth is too massively complex to answer a most primary question, it did not stop the Supreme Court of the United States of America from ruling that global warming is real and carbon dioxide is causing it.

GUILTY

Despite insufficient evidence to support claims of anthropogenic global warming, those who stubbornly cling to the failed hypothesis scored a monumental victory on April 2, 2007. Science was thrown under the bus by the Supreme Court when five black-robed judges declared that carbon dioxide was a *pollutant*. The petitioners were the states of California, Connecticut, Illinois, Maine, Massachusetts, New Jersey, New Mexico, New York, Oregon, Rhode Island, Vermont, and Washington, as well as the cities of New York, Baltimore, and Washington, D.C., the territory of American Samoa, and radical organizations including Center for Biological Diversity, Friends of the Earth, Greenpeace, Natural Resources Defense Council, and the Sierra Club.

The Court's landmark 5-4 decision required the Environmental Protection Agency to regulate CO_2 under the Federal Clean Air Act. President Bush (God bless his efforts on this one) refused to allow his EPA to respond to the court ruling, but within his first month on the job, predictably, Barack Obama did.

December 7, 2009, lived up to its 1941 billing as the day that will live in infamy, as Obama's EPA chief Lisa Jackson held a press conference to announce "that the EPA has finalized its endangerment finding on greenhouse gas pollution, and is now authorized and obligated to take reasonable efforts to reduce greenhouse pollutants under the Clean Air Act."

The "endangerment finding" means humanity is at risk because of the miniscule increase in carbon dioxide, and the Obama administration believes it has the dutiful power to react by curtailing such emissions.

"The overwhelming amounts of scientific study show that the threat is real—as does the evidence before our very eyes," Jackson said. "Polar ice caps crumbling into the oceans, changing migratory patterns of animals and broader ranges for deadly diseases, historic droughts, more powerful storms, and disappearing coastlines. After decades of this mounting evidence, climate change has now become a household issue. Parents across the United States and around the world are concerned for their children and grandchildren."

Taken to an extreme, virtually anything that emits CO_2 could be regulated by the EPA—power plants, planes, cars, lawnmowers, even fizzy drinks and marathon runners.

The Clean Air Act was signed into law by President Richard Nixon in 1970. It resulted in a major shift in the federal government's role in air pollution control. The legislation authorized the development of comprehensive federal and state regulations to limit emissions from both stationary (industrial) sources and mobile (vehicle) sources. To implement the enforcement of these new laws (and to spinelessly cater to the anti-war, eco-freak hippies), "Tricky Dick," as the president proved himself to be, created the Environmental Protection Agency in 1971. He did so by Executive Order—in other words there was no vote by the Congress for or against the measure. Today, the EPA employs over 18,000 people and gobbles up an annual budget of $9 billion.

It was critical for the petitioners addressing the Supreme Court in 2007 to convince the judges that carbon dioxide was a pollutant, because the Clean Air Act had only allowed the EPA to regulate pollution—not natural fertilizers.

And if Tricky Dick was able to create an entire government agency with the stroke of a pen, the Obama administration certainly can regulate carbon dioxide and other so-called greenhouse gases without a vote of Congress.

I've never been a big fan of the EPA—it's a division of the federal government that is regularly used to curtail free enterprise and usurp private property. Now, with its new responsibilities, the EPA is becoming the new KGB. Through the counterfeit regulation of carbon emissions, the federal government is poised to level the playing field amongst the classes, thereby more efficiently controlling our lives.

Meantime, a few people, like Al Gore, stand to make *billions* off of these government shenanigans.

5

GORE ON THE GRILL

If you think it's about greed, you don't know me . . .

—Al Gore before Congress, April 24, 2009

I HAD JUST concluded a radio broadcast when my cell phone began buzzing. It was an acquaintance from a charitable organization I am involved in. It was a bit of a surprise to hear from this man, because other than a few brief conversations, I really didn't know a whole lot about the guy.

"You were talking about Al Gore's association with a Silicon Valley venture capital firm and said you thought he stood to make millions off of global warming."

"Yeah, that's what I said." I had no idea where this conversation was going.

"Well, you were incorrect. I'm intimate with the venture group you were referring to."

"Well, what did I get wrong?"

"For starters, Gore stands to make *billions*, not millions—which he's already started raking in."

I immediately arranged for a meeting.

FRIENDS AND FAMILY

Albert Gore, Jr., was welcomed into the world with high expectations. His father was an influential Tennessee U.S. Representative and Senator who served in Congress from 1939 to 1971. During that period Gore Sr.'s net worth amazingly swelled from middle class to upper crust. By the time "Little Al" (as Gore Sr. nicknamed his firstborn son) was being raised, the Gore family was living most of the year in Suite 809 in Washington, D.C.'s luxurious Fairfax Hotel. Directly below Little Al's suite lived the powerful Senator from Arkansas, John McClellan, and above, an extraordinarily wealthy widow named Madame Brambilla, who rented an additional suite just to keep her extensive wardrobe; she also owned another residence in Rome.[105] Surrounded by the rich and famous, Little Al is said to have told his daddy, "One day, I'm going to be somebody."[106] After all, daddy had made himself a "somebody," thanks in part to a Communist enabler by the name of Armand Hammer, chairman of Occidental Petroleum Company.

In the run-up to the 2000 presidential election, when Gore Jr. came as close as one could to becoming commander-in-chief, many investigative journalists began reporting the eye-opening story behind the Gore's family friend. Perhaps the most damning work flowed from the pen of Edward Jay Epstein in his 1996 book, *Dossier: The Secret History of Armand Hammer*. Drawing from FBI and Russian intelligence documents, Epstein illustrated Hammer's extensive business dealings with Russia and the former Soviet Union, dating back to the reign of the murderous Josef Stalin. Armand Hammer laundered millions of dollars for the Soviet Union in bogus business transactions, mined asbestos, and brokered the production of tractors and even pencils for Stalinist Russia. He traded fur and trafficked in Czarist art, both real and phony. At one point, in a bizarre exchange of communist conversation, Vladimir Lenin told Stalin that Hammer was a "path leading to the American business world, and this path should be made use of in every

---

105 Maraniss, David and Nakashima, Ellen, "Al Gore: Growing Up in Two Worlds," *Washington Post*, Sunday, October 10, 1999, A1.
106 Ibid.

way."[107] Later, Epstein revealed that through a combination of shrewd dealing and bribery, Hammer acquired Libyan oil rights for Occidental Petroleum.

A June 2000, *Wall Street Journal* story revealed how Gore Sr. had gained much of his wealth with Hammer's aid.[108] First, there was a sweetheart deal to garner a substantial profit in the cattle breeding industry, followed by a low-cost land sale, complete with mineral mining rights to a known zinc mine underground, and, eventually, a directorship on the board of an Occidental subsidiary, Island Creek Coal Company—a position reportedly worth $500,000 annually. The director's seat was an apparent payback for papa Gore's help in trying to secure a long-term lease on a chemical manufacturing plant in West Virginia, owned by the U.S. Army. Hammer had petitioned the government for a license to export synthetic nitrogen and ammonia-based fertilizer (the kind that can be used as a key ingredient in the making of bombs). He wanted to use the plant as his base of operations. Senator Gore was instrumental in trying to shepherd the plan through government channels, but fortunately, the deal was stopped by the House Armed Services Committee.

In 1972, Armand Hammer pleaded guilty to three counts of making illegal campaign contributions—$54,000 in $100 bills—to a Nixon fundraiser. Apparently, his hope was that Nixon would be the man to normalize relations with the Soviet Union, a move that would greatly aid Hammer financially.

The Gore/Hammer story becomes even more disturbing when Hammer's family history is revealed. His father, Julius Hammer, was born in Russia in 1874. At 16, he moved to New York with his family. Julius attended medical school at Columbia College, became a doctor, and eventually gained a seedy reputation as an abortionist. In addition, he ran eight drugstores with proceeds reported to have been sent to support the Bolshevik Party in Mother Russia—a party which included the likes of Vladimir Lenin, Leon Trotsky, and Josef Stalin. Not surprisingly, Julius became one of the early leaders of the Communist Party in America.

---

107 Morrison, Micah, "Al Gore Jr.: Occidental Oriental Connection," *Wall Street Journal*, June 29, 2000.
108 Ibid.

In 1898, Armand was born into this nest of anti-American values in Manhattan, eventually continuing the relationship with communist Russia that his father had established. And, just as Armand Hammer helped Albert Gore Sr. become somebody, he would later do the same for Little Al.

HIGHER EDUCATION

After attending the prestigious St. Albans School in D.C., Gore, Jr. went on to Harvard where his school experience was lackluster, to say the least, including a D in a course that one would think would have been a slam-dunk for the future Nobel Laureate: "Man's Place in Nature."[109] Despite later claims that "during my service in the United States Congress, I took the initiative in creating the Internet,"[110] at Harvard, Al steered clear of all courses in mathematics and logic. According to the *Washington Post*, fellow students "remember him spending a notable amount of time in the Dunster House basement lounge shooting pool, watching television, eating hamburgers, and occasionally smoking marijuana."[111] At Harvard, he also met his future bride, Tipper. According to Al, it was a love story made for the silver screen.

In November 1997, *Time* magazine reported that a chummy Vice President Al Gore was talking to reporters riding in the press compartment of his plane, Air Force Two: "Sliding behind a table with the two reporters covering him that day, he picked slices of fruit from their plates and spent two hours swapping opinions about movies and telling stories about old chums like Erich Segal, who, Gore said, used Al and Tipper as models for the uptight preppy and his free-spirited girlfriend in his book *Love Story*."

Apparently, no one ever asked Tipper how she felt about this, especially given that the free-spirited girlfriend in the film tragically dies.

---

109 David Maraniss and Ellen Nakashima, "Gore's Grades Belie Image of Studiousness," *Washington Post*, Sunday, March 19, 2000, A1.
110 CNN *Late Edition* with Wolf Blitzer, March 9, 1999.
111 David Maraniss and Ellen Nakashima, "Gore's Grades Belie Image of Studiousness," *Washington Post*, Sunday, March 19, 2000, A1.

Segal was reportedly so ticked at Gore's claim that according to the *New York Times* he contacted both the press and Gore and told them he was "befuddled" by the *Time* story.[112] Gore's responded by claiming the whole thing was just a misunderstanding.

Setting the record straight, Segal said he did meet Al Gore while teaching at Harvard in 1968. However, the male lead in *Love Story* was primarily based on the personality of another future star, actor Tommy Lee Jones. Al Gore's controlling father, Segal said, provided the inspiration for the character's drive to follow in daddy's footsteps, and the female lead was in no way based on Tipper. In fact, Segal made it clear to the *Times* that the romance in the novel and film was not based on the couple at all. "I did not draw a thing from Tipper," he said. "I knew her only as Al's date."[113]

Why spend time on Al Gore stories from the Sixties? Because it appears Gore has a wild track record of imaginatively embellishing anything that might serve him well. Perhaps it was the dope at Dunster, or perhaps it's just a deeply engrained need to be "somebody."

Besides the poor grades, munchies, and Tipper, we also know a bit about Al's political leanings while at Harvard which were far left of center and seemingly in line with the ties his father had forged with Armand Hammer. It is reported that in a letter sent to his Senator dad, Gore commented on America's fear of Communism, "We do have inveterate antipathy for communism—or paranoia, as I like to put it." He went on to allegedly say that such a "psychological ailment" was a "case of national madness."[114]

Given that communists like Hammer's associate Stalin killed an estimated 43 million people to achieve his political objectives, one would *expect* most Americans to hold to a healthy dose of paranoia regarding the evils of communism, but apparently not so with Al Gore.

Following graduation from Harvard with a Bachelor of Arts degree, Gore enlisted in the Army, was trained as a journalist, served five months as a military reporter in Vietnam, and eventually obtained an early release from his two-year commitment.

---

112 Melinda Henneberger, "Author of 'Love Story' Disputes a Gore Story," *New York Times*, December 14, 1997.

113 Ibid.

114 Roger Banks, "The Education of Al Gore," *Washington Times*, March 25, 2000.

Coming home, he secured a job at the *Tennessee Reporter* newspaper, married Tipper, and, perhaps trying to find himself, enrolled in the Vanderbilt University Divinity School. But learning about God did not work out so well for the future environmental savior. In the eight classes he took over three semesters, he apparently received five F's and then dropped out.[115]

Proving no shame, in 1999 Gore was invited to the Vanderbilt Divinity School where he spoke to the students. Undaunted, he decreed, "It's great to be back here. This was one of the most important years I ever spent in my life. I had wanted to come here after returning from Vietnam to have a structured opportunity to explore the most important questions in my life . . . I've frequently drawn on the lessons that I learned back then."[116]

What "lessons" could Gore have been referring to? Bombing out of the study of God should be a tell-tale sign that the Almighty wasn't really on his agenda. But his next steps would seem to indicate fame and the almighty dollar certainly were.

Having flunked out of divinity school, Gore took a brief shot at a Vanderbilt Law degree. However, when Tennessee's Fourth District Congressional seat—once held by his father—unexpectedly opened up in 1976, Gore dropped out of law school and hastily made his first run for public office. He managed to squeak through the primary as a pro-life Democrat and, with obvious name recognition, ran unopposed to victory in the fall. Finally, he was on track to becoming somebody. In 1984, he would further assume the mantle of dear old dad, representing Tennessee as a prochoice Democrat in the U.S. Senate. Gore would also continue the relationship with Gore Sr.'s communist friend.

HAMMER TIME

By the time Al Gore Jr. was elected to Congress, his father had been out of office for six years. Shortly after his defeat to William Brock in 1970, Larry Bates, a member of the Tennessee legislature, is reported to

---

115 Ibid.
116 Lew Harris, "Campus Visit Sparks Al Gore Memories of VU Years," *Vanderbilt Register*, August 2, 1999.

have flown with Gore Sr. from Nashville to Washington, D.C. "Since the voters don't want me anymore," Gore said, "I'm going to take my expertise and make some serious money for myself, my good friend Armand Hammer, and Occidental Petroleum."[117]

Among his first tasks with Occidental was renegotiating coal contracts between Occidental subsidiary Island Creek Coal and the Tennessee Valley Authority. Gore Sr. became the director of Island Creek Coal in 1972. In a sweetheart land deal, Gore Sr. reportedly paid $80,000 for a large parcel of land near his estate on the Caney Fork River. Having inside knowledge of the property's mineral assets, he also paid another $80,000 for the rights to mine the property for zinc. The former senator eventually sold the entire package to his son Al for no profit. Then, in a move that Little Al would replicate later in life, Occidental paid zinc royalties totaling $20,000 a year to the future environmental leader. Mining operations began on the property in 1985 with annual payments reportedly flowing into Al Jr.'s pockets until 2003.[118]

As soon as Al ran for Congress in 1976, Armand Hammer became a regular, big-time campaign contributor. Federal Election Commission records show that Hammer, his wife, his corporations, and family members provided legal contributions to Al's campaigns.[119] Hammer's former personal assistant, Neil Lyndon, has publicly stated that Hammer, Al, and Tipper regularly dined together, and sometimes Hammer would even allow the couple to use his luxurious private Boeing 727 for "their own journeys and jaunts."[120]

In 1987, Senator Al Gore and Hammer flew to Moscow, Russia, the heart of communism, where Hammer was to receive a humanitarian award from the International Physicians Against Nuclear War. Gore also spoke to the 2,000 communists and commie sympathizers in attendance, advocating nuclear disarmament.

---

117 Charles C. Thompson II, and Tony Hays, "CIA Official: 'Gore Compromised by Secret Past,'" WorldNetDaily.com, October 16, 2000.
118 Charles C. Thompson II and Tony Hays, "Al Gore, polluter?" WorldNetDaily.com, October 20, 2000.
119 Neil Lyndon, "How Mr. Clean Got His Hands Dirty," *Sunday Telegraph*, November 1, 1998.
120 Ibid.

In the summer of 1992, Al Gore's bestseller, *Earth in the Balance*, was published, knighting him the world's foremost environmentalist and establishing him as global warming's go-to guy. That fall he was elected Bill Clinton's vice president. Despite his green credentials, numerous reports indicate that he continued to do huge favors for Occidental Petroleum (Hammer died in 1990) despite the zinc royalties and his father's status as a prominent Occidental shareholder (his dad died in 1998).

Perhaps the most egregious example of Gore's hypocrisy was the oilfield deal the Clinton Administration brokered near Bakersfield, California. The Elk Hills oilfield had been set aside decades ago as a strategic military reserve in the desert outside of Bakersfield. Ever since the 1920s, scandalous efforts by politicians had been made to sell the reserves to private companies. In 1921, President Warren Harding's Interior Secretary Albert Fall went to jail for taking a $100,000 bribe to approve a lease of the field to the Pan American Petroleum Company in what would be known as the Teapot Dome Scandal. For the next 70 years, the stain of Teapot Dome remained vivid enough to halt attempts to privatize the military's reserves of black gold at Elk Hills. Even during the Nixon and Reagan administrations, when attempts were made to sell the reserves, Congress resoundingly said "no."

However, this all changed during the green Clinton-Gore reign. A provision was neatly tucked into the massive 1996 Defense Appropriations Bill that finally allowed for the sale of Elk Hills to private interests. Oil companies salivated at the opportunity to bid for this long-sought-after land. And guess what company was able to purchase this prize real estate? You guessed it—Occidental Petroleum. Occidental purchased the majority of Elk Hills for $3.65 billion in 1998. It was the largest privatization deal in United States history, and it tripled Occidental's domestic reserves.

Connecting the dots between Armand Hammer, Albert Gore Sr., and Little Al, many journalists tried to review the documents relating to the sale, but then-Department of Energy Secretary Bill Richardson (the same guy who was a Democratic presidential candidate in 2008, and the same guy who had to bow out from being named Obama's secretary of commerce because of pay-for-play allegations as Governor of New Mexico) refused to release the information.

To this day, Gore's surrogates claim that his hands are clean of any wrongdoing. They will tell you that even though he inherited his dad's immense share of Occidental stock, he divested himself of it in 2000. They will also remind you that he stopped receiving royalties from Occidental subsidiary Island Creek Coal in 2003. However, most can see straight through this charade. Throughout the Nineties Al Gore was an environmental hero leading a double life replete with monthly mineral money running into his personal coffers and a large, oil-based inheritance in his personal pipeline.

Regarding the glaring hypocrisy, while campaigning for president in 2000, Gore fed skeptical reporters in Tennessee his story about the Occidental money inherited from daddy: "According to his will, that was put into a trust fund to benefit my mother for the remainder of her life, and I was named executor of his will. I have legal obligations as executor which are very stringent. I have to comply with the provisions of the will. . . . If you can find anything wrong with that, please tell me."[121]

Asked by the same gaggle of reporters if there was anything wrong with the lengthy relationship between his family and Armand Hammer's Occidental Petroleum, Gore arrogantly replied, "There certainly is not."

Maybe the relationship with Hammer is not a big deal to Gore, but to the rest of us, it's just one of many big deals.

AL'S INSPIRATION?

If you were to ask Al Gore who is the chief denier of man-caused global warming, he would likely answer that it was Dr. Fred Singer. In 2006, Singer and Gore delivered separate addresses to the American Geophysical Union's annual convention in San Francisco. Singer had been strategically placed on the schedule just prior to Gore's headlining speech. A day prior to the event, Dr. Singer joined me for a live radio interview, and after the interview I gave him a ride back to his hotel. It was certainly an honor for me, because ever since my days in TV weather, I knew Fred Singer as one of the pioneers of modern atmospheric science.

---

121 Karin Miller, "Gore Defends Occidental Link," Associated Press, March 14, 2000.

Dr. Singer's *Curriculum Vitae* is overwhelmingly impressive, beginning with his degree in electrical engineering from Ohio State and his Ph.D. in physics from Princeton; pioneering the development of rocket and satellite technology; having published more than 400 technical papers in scientific, economic, and public policy journals; as well as his numerous editorial essays and articles in the *Wall Street Journal, New York Times, Newsweek, Washington Times, Washington Post*, and other publications. He was also the nation's first director of the National Weather Satellite Service, and currently serves as president of the wonderful group he founded, the Science and Environmental Policy Project (SEPP.org). Simply stated, the man is a genius.

Singer looks and sounds the part of a Virginia gentleman, and, despite his 80-plus years, he is full of vigor.

Riding high in my large, four-wheel drive, extended-cab truck, I pulled up to the curb in front of the station and rolled down the tinted window.

"Dr. Singer, your car, sir," I said, above the purr of the idling engine.

"Brian, this will be a first. I don't think I've ever been in a vehicle this big."

"Don't worry, Doctor. With all the carbon this truck is pumpin' out, the few trees that grow in this part of 'Sodom by the Sea' will love us."

Singer chuckled and climbed aboard. He asked if there was a copy center nearby where he could print some slides for his talk at the convention the next day. It was fine with me; we would have additional time to talk. I asked him if he was excited to be speaking prior to the former vice president.

"Oh, yes. And I'm going to take the opportunity to encourage my colleagues to dare challenge Mr. Gore on some key issues."

"Specifically?"

"Specifically, his predictions on sea-level rise and a few other key issues from his latest book."

"You know, Doctor," I said with a grin, "the condo Gore owns in San Francisco will be underwater if his predictions come true. Doesn't sound like a prudent investment to me."

"I'll make a note of that," he said, smiling.

"Will you get into the Revelle story at all? It's really quite a whopper."

"I won a lawsuit over that, Brian. There's no need to belabor the issue."

"But Gore still speaks of the enlightenment he received at the feet of Dr. Revelle."

"I have something on the subject in my briefcase. I'll make a copy for you. It will provide you with all the details from the lawsuit."

In virtually all of Gore's media, he is quick to give praise to Dr. Roger Revelle, an outstanding oceanographer who was a visiting professor at Harvard while Gore was a student there. Gore has repeatedly claimed that Revelle acted as an academic mentor to him, a curious thought since Gore was never noted for his academic pursuits.

Singer and Revelle first met when both were young scientists in 1957. Revelle was conducting groundbreaking measurements of atmospheric carbon dioxide that led him and colleagues at the Scripps Institute of Oceanography near San Diego to conclude that not all carbon dioxide emitted from the burning of fossil fuels would be quickly transferred to the ocean. Some of the CO_2, he theorized, might accumulate in the atmosphere. Singer, like others at the time, was very intrigued by the theory.

Over the years, Singer and Revelle had forged a significant professional and personal friendship. In 1968 Revelle participated in an American Association for the Advancement of Science (AAAS) symposium organized by Dr. Singer, regarding the possible effects of CO_2 on climate. In the Seventies, while teaching at Harvard, Revelle invited Singer to spend a few months as a faculty guest. During that period, for a few weeks, Singer was Revelle's personal houseguest.

Years later, in February 1990, at the annual meeting of the AAAS in New Orleans, Revelle and Singer had a breakfast meeting in the hotel restaurant. Global warming, or "greenhouse warming" as it was then called, was gaining a certain amount of media attention, and the two decided to write a paper about the greenhouse effect and its potential effect on planetary warming. While working on the article, the two decided to include Dr. Chauncey Starr as a coauthor because of his expertise in energy research. The article was submitted to the Cosmos Club of Washington, D.C., and published in the *Cosmos* journal. "What to Do About Greenhouse Warming: Look Before You Leap" appeared in the April 1991 edition.

Singer told me the conclusion to their article was simple: "The scientific basis for greenhouse warming is too uncertain to justify drastic action at this time. There is little risk in delaying policy responses." This conclusion was very similar to a 1988 warning issued by Revelle in a letter he sent to members of Congress after James Hansen's wild Congressional testimony of global gloom and doom.

Three months after the *Cosmos* article appeared, Revelle passed away.

About a year later, Gore's *Earth in the Balance* hit the stores. In a *New Republic* article entitled "Green Cassandras," journalist Gregg Easterbrook made reference to the Singer-Revelle-Starr paper. Easterbrook reported that despite Gore's claim in his book that Revelle was his inspirational mentor, there is no mention in *Balance* that Revelle even contributed to the *Cosmos* article questioning global warming.

The mainstream media began to pick up on the discrepancy, and Gore's green credibility was threatened.

"In July, 1992," Singer told me, "I received a phone call from a Dr. Justin Lancaster, who introduced himself as being from the Environmental Science and Policy Institute at Harvard and a former associate of Roger Revelle."

Almost immediately, Singer said, the conversation took an odd turn. Lancaster requested that Revelle's name be removed from the greenhouse paper published in *Cosmos*. Singer refused, although technically, *Cosmos* held the copyright.

Gore was soon hit with the greenhouse controversy in a vice presidential debate with Dan Quayle and James Stockdale. I vividly recall watching this debate on television in October 1992. For a brief, and for some, pleasurable, moment, Gore was made to look like a total moron—complete with howling laughter from the live audience. The debate transcript reads:

Stockdale: I read where Senator Gore's mentor had a disagreement with some of the scientific data that is in his book. How do you respond to those criticisms of that sort? Do you ... take this into account?

Gore: No, I—let me respond. Thank you, Admiral, for saying that. You're talking about Roger Revelle. His family wrote a lengthy letter saying how terribly he had been misquoted and had his remarks taken completely out of context just before he died.[122]

---

122 The Gore-Quayle-Stockdale Vice Presidential Debate, October 13, 1992, Commission on Presidential Debates.

At this point the transcript shows—and I remember—that the audience began to jeer Gore's response, causing Gore to look terribly uncomfortable. He continued his response, trying desperately to remain cool above the audience's overt skepticism.

Gore: He believed up until the day he died—no, it's true, he died last year . . .

The official transcript notes that again the humiliation continued with more audience jeers and laughter.

Moderator: I'd ask the audience to stop, please.

Gore: . . . and just before he died, he co-authored an article which was—had statements taken completely out of context. In fact, the vast majority of the world's scientists—and they have worked on this extensively—believe that we must have an effort to face up to the problems we face with the environment. And if we just stick our heads in the sand and pretend that it's not real, we're not doing ourselves a favor. Even worse than that, we're telling our children and all future generations that we weren't willing to face up to this obligation.[123]

I asked Singer, "The activists really decided to go after you following the vice presidential debate, didn't they?"

"Well, that's when they certainly became aggressive," he replied.

A memorial symposium in Revelle's name was held at Harvard two weeks after the debate. In written remarks for the event, Lancaster went on record suggesting that Roger Revelle had never been a co-author of the *Cosmos* article and that Singer added Revelle's name over his objections.

"He later added the charge that I had pressured an aging and sick colleague, suggesting that Dr. Revelle's mental capacities were failing at the time," Singer told me.

---

123 Ibid.

Hearing that the now-famous *Cosmos* article was about to be republished, a member of Gore's staff, Dr. Anthony Socci, sent a letter to the publisher requesting that the article be dropped.

Tired of the defamation of his character, a libel lawsuit was filed by the Center for Individual Rights on behalf of Dr. Singer in 1993. During the discovery phase of the trial, it was learned that Lancaster had been phoned personally by Al Gore after the "Green Cassandras" article appeared. Gore wanted to know if Revelle's mental faculties were sharp enough to properly reflect his views in *Cosmos*.

"In a draft of a reply letter to the senator, Dr. Lancaster completely undermined the claims he later made against me, stating Dr. Revelle was 'mentally sharp to the end,'" Singer said.

That letter was found on a computer disk belonging to Lancaster. Revelle's personal assistant also provided the court with documents proving that during the period that Gore seemed to be hoping his mentor was mentally incapable of collaborating on the *Cosmos* article, he was perfectly fit. Other court records indicated that Revelle had been maintaining a full schedule during that period, including traveling and speaking. So, the simple truth was that Dr. Revelle knew exactly what he was writing.

In addition, Lancaster also wrote[124] that Revelle had shown him the *Cosmos* manuscript prior to it being published, with the comment that he "felt it was honest to admit the uncertainties about greenhouse warming, including the idea that our ignorance could be hiding benefits as well as catastrophes." Lancaster stated that Revelle had "agreed there did not seem to be anything in the article that was not true."

Dr. Singer won his libel suit, and in 1994, Dr. Lancaster wrote him a lengthy letter of apology.

SORE GORE

On the February 24, 1994, edition of ABC's *Nightline*, host Ted Koppel revealed that Vice President Al Gore had called him and suggested that

---

124 J. Lancaster, draft letter to Senator Al Gore, July 20, 2002, found on a computer disc belonging to Lancaster during the discovery process.

he investigate the political and economic motivations of the "antienvironment" movement—specifically, vocal skeptics of global warming, most likely as in Dr. Fred Singer. Koppel refused to be bullied by Gore, and instead stated, on the air, "There is some irony in the fact that Vice President Gore—one of the most scientifically literate men to sit in the White House in this century—that he is resorting to political means to achieve what should ultimately be resolved on a purely scientific basis."

This page from Gore's 1994 playbook is currently the same tactic used by global warming zealots today: attack and discredit skeptics and deniers of global warming based on funding, education, religious and political affiliation, or, when in a jam, simply by characterizing them with humiliating brands.

Such was the case when Gore appeared on NBC's *Today Show* on November 5, 2007. Host Meredith Vieira quoted from an op-ed that ran a few days earlier in the *Wall Street Journal*. The author of the piece, John Christy, a member of the United Nation's Intergovernmental Panel of Climate Change, stated, "I see neither the developing catastrophe nor the smoking gun proving that human activity is to blame for most of the warming we see."[125]

Gore informed Vieira, "Well, he's an outlier. He no longer belongs to the IPCC. And he is way outside the scientific consensus."

Dr. John Christy, an "outlier?" Christy is director of the aforementioned Earth System Science Center at the University of Alabama in Huntsville, where they monitor the earth's temperature via satellite. He was also a respected participant in the U.N.'s Intergovernmental Panel on Climate Change, and as such, *a co-recipient of Al Gore's Nobel Peace Prize*!

Gore then fired a warning shot across the news media's bow, piously informing Vieira that "part of the challenge the news media has had in covering this story is the old habit of taking the 'on the one hand, on the other hand' approach. There are still people who believe that the earth is flat. But when you're reporting on a story like the one you're covering today, where you have people all around the world, you don't search out for someone who still believes the earth is flat and give them equal time."

---

125 Dr. John R. Christy, "My Nobel Moment," *The Wall Street Journal*, November 1, 2007.

UNGREEN GORE

Early in 2007, while rumors began circulating that Al's film, *An Inconvenient Truth*, was about to win an Oscar and that he was a slam-dunk for the Nobel Prize, the Tennessee Center for Policy Research was awarding Gore with "a gold statue for hypocrisy."[126]

The Center revealed that Gore's mansion in Nashville's upscale Belle Meade neighborhood consumes more electricity in one month than most of us use in a year—with some months consuming *twice* the national annual average.

According to statistics from the Department of Energy, the average household in America utilizes 10,656 kilowatt-hours of electricity per year. In 2006, Gore's house devoured 221,000 kilowatt-hours. In August of that year, he apparently ripped through 22,619 kilowatt-hours. The Center said Gore paid nearly $30,000 in combined electricity and natural gas bills to power his Nashville house in 2006.

After the Tennessee Center for Policy Research exposed Gore's massive home energy use, the former vice president scurried to make his home more energy-efficient. Despite adding solar panels, installing a geothermal system, replacing existing light bulbs with more efficient ones, and overhauling the home's windows and ductwork, apparently Gore's home now consumes *more* electricity than before the "green" overhaul. According to the Center:

> Since taking steps to make his home more environmentally-friendly last June, Gore devours an average of 17,768 kWh per month—1,638 kWh more energy per month than the year before the renovations. By comparison, the average American household consumes 11,040 kWh in an entire year, according to the Energy Information Administration. The cost of Gore's electric bills over the past year topped $16,533.[127]

---

126 "Al Gore's Personal Energy Use Is His Own "Inconvenient Truth," Press Release, Tennessee Center for Policy Research, February 26, 2007.
127 "Al Gore's Personal Electricity Consumption Up 10% Despite 'Energy Efficient' Renovations," Press Release, Tennessee Center for Policy Research, June 23, 2008.

Hopefully, Gore won't be involved in further green renovations—they only seem to expand his enormous carbon footprint. This, of course, is not the only home Gore owns. In addition to his 10,000 square-foot Nashville estate, Al and Tipper also have a spacious home in Arlington, Virginia, a third home adjacent the zinc field in Carthage, Tennessee, and a posh multimillion dollar condo at the St. Regis in San Francisco that boasts amenities such as 24-hour room service, and a full-time butler that, according the building's website, "brings any request and whim to brilliant realization." There's nothing on the website however, about being green.

VERY GREEN GORE

It's widely reported that Al Gore is worth at least $100 million, although my well-connected acquaintance believes it may be closer to $500 million.[128] Quite a success story for a guy, whom, according to financial disclosure records released just prior to his bid for the presidency, had a net worth near $2 million.

Besides Gore's best-selling books, his Oscar-winning film, and the steady flow of six-figure speaking gigs, Al Gore is a huge player in the Silicon Valley. He has been a senior advisor to Google since before the company went public in 2001—a deal which provided him with stock options estimated at $30 million, and he joined the Apple board in 2003, receiving millions more in stock, plus a handsome annual salary. Google has been pushing a variety of green government policy strategies, and Apple has pushed a green marketing campaign. Both stand to benefit from Gore's continual global warming charades.

In 2004 Gore co-founded the London-based hedge fund, Generation Investment Management (GIM). GIM appears to be heavily involved in carbon trading. It also appears he is set to trade carbon credits in the United States, too. Again, whenever Gore blows his horn to save the world from the evils of carbon dioxide, his personal slot machine likely hits the jackpot. In 2007 Gore joined the board of arguably the most powerful

---

128 Ellen McGirt, "Al Gore's $100 Million Make-Over" *Fast Company* magazine, December 19, 2007, http://www.fastcompany.com/magazine/117/features-gore.html.

venture capital firm in the Silicon Valley, Kleiner Perkins, Caufield & Byers (KP)—the same group that took Google public. According to my personal mole, the firm had already invested nearly a billion dollars into "green" start-up companies.[129]

"Okay," I asked my acquaintance, who, by the way, has been in closed door meetings with Gore, "does Al Gore really believe that the earth has a fever, and he has the remedy to fix it?"

Slowly shaking his head from side to side, he replied, "Gore believes in money, as I will illustrate."

And tragically, while Al Gore and his fellow elites cash in on global warming, the drones viewing his media presentations are buying the gloom and doom as truth, rather than the fiction it is.

---

129 Confirmed in a story by the *New York Times* "Gore's Duel Role: Advocate and Investor," John M. Broder, November 2, 2009.

6

CONVENIENT LIES

Communists must always consider that of all the arts the motion picture is the most important.

—Vladimir Lenin

MY FAMILY AND I were sharing an extended weekend with some wonderful friends we'd invited to our cabin on Donner Lake, near Truckee, California. Sledding was on the day's menu, and our friends suggested a nearby sled park they had visited some years ago, owned and operated by the Sierra Club.

The *Sierra Club?* I asked, feigning shock. "They own a *sledding park?* With all the imminent global warming they preach, I'm surprised they haven't converted it to a year-round skateboard park."

"Or maybe a swim center and outdoor tanning salon," my buddy Daniel quipped, catching my sarcastic drift.

"No," his wife Laurie said. "They really run a sledding park. It's right next to the hostel they own. Loads of people stay there every night."

"Loads of people, or loaded people?" I couldn't resist.

"This'll be great," Daniel said. "Mr. No-Such-Thing-As-Global-Warming is going to a sledding park-slash-hippy hotel, owned by the Sierra Club."

We packed the kids and sleds into two four-wheel drive SUVs and headed for the park. Twenty minutes later we arrived at the Clair Tappaan Lodge.

"Is it just me, or are you guys sensing the irony here?" I asked as I observed our surroundings. The parking lot was packed with a variety of gas-guzzling SUVs, most boasting Sierra Club bumper stickers.

"The bumper stickers offset the SUV's carbon footprint," Daniel joked.

"Right on, bro'."

Before sledding, a nominal fee was to be paid inside the hostel. We trudged up the snow-covered walkway to the lodge entrance. Laying our sleds aside we lined up to proceed through the bulky, wooden entry door, fashioned with rustic, heavy duty hardware.

Quite a manly door, I thought to myself, pushing it open.

That is, manly until—

"Al Gore!" my wife exclaimed, startled.

"What the—," I said.

There he was—looking straight at us, his life-size face plastered on a poster tacked to the interior wall facing the backside of the door.

The poster read:

an inconvenient truth
the crisis of global warming
AL GORE
*LEARN WHAT **YOU** CAN DO.*

Scrawled in a marker at the base of the poster was an announcement that Gore's Oscar winning film, *An Inconvenient Truth*, would be shown that evening at 8 o'clock in the cafeteria. Discussion would follow.

Mr. "No-Such-Thing-As-Global-Warming" was walking straight into the belly of the beast.

"Oh, this is going to be fun," Daniel said, as everyone laughed.

Perhaps it was fun for Daniel, our wives and children, but once inside the lodge, I became perturbed. Beneath the big poster was a table brimming with Al Gore pamphlets and take-home posters explaining how "you can do your part to stop global warming." Similar propaganda filled the wood-paneled hallways of this otherwise charming retreat. It was like entering the headquarters of a damned religious cult. I wanted to investigate further.

I asked my wife, "Could you ladies pay for us while Daniel and I snoop around?"

We shuffled up a staircase and soon found ourselves in the library. It was the size of a large master bedroom, with a variety of cozy sitting areas. A fire flickered behind the glass door of a glowing wood stove. A smattering of adults sat in the well-worn furniture, engrossed in quiet reading.

The walls were covered with shelves holding books that were predictably left of center. Besides the Sierra Club's nature publications, there were copies of communist sympathizer Noam Chomsky's *America's Quest for Global Dominance*, atheist Sartre's metaphysical missive *Being and Nothingness*, and the obligatory books by Barack Obama. However, the most prominent author on the shelves was Al Gore. Multiple copies of everything written in his name were available, not just on the shelves, but on anything that was flat, including—spread out like the family Bible— the oversized coffee table version of *An Inconvenient Truth*; complete with compelling photographs, oversimplified charts and uncorroborated statements regarding a planet that was near disrepair. A special display table was set up with pamphlets revealing how to "save the planet in your daily life" and "how you can become a trained volunteer to share Gore's message in your community." And, of course, there was the poster—Gore's movie would be shown *tonight at 8*.

"This place is a lie-brary," I said aloud, with precise enunciation.

Heads turned to look at me, but I didn't really care at that point.

"Let's hit the sled run," I told Daniel.

ABSORBING THE LIES

That night, worn out from our day of climbing up and down the sledding hill with the kids, Daniel and I sat next to a crackling fire burning in our fireplace. Our hearts content with food and beverage, we began to muse about our adventure in the halls of the hostel.

"Didn't you take one of those posters on how to save the earth for me?"

"Yeah. It's right over here."

Daniel fished out the poster from a pile of his personal reading materials and handed it to me. It was the same one we had seen pinned up all over the lodge. The front of the poster promoted *An Inconvenient Truth*. On the backside was a how-to guide on *Taking Action!* The poster was creatively produced with folds, so that once reduced, the how-to side opened like a book, providing the reader with "save the climate" talking points.

"Let's walk through this garbage together," I suggested.

"All right. I'm the student," Daniel said. "Go."

"'Activity One: Understanding the Greenhouse Effect. Have students create their own greenhouse conditions. Put a nail hole in a closed plastic bottle. Insert the tip of a thermometer into the hole. Place bottle in sun. Place a second thermometer next to the bottle. Watch the temperature of the bottle warm higher than the air outside it.'"

"And this proves what?" Daniel asked.

"It says this will encourage students to 'consider how the bottle acts similarly to greenhouse gases by trapping heat.'"

"So, I guess the idea is, humans are producing the greenhouse gases that warm the earth?"

"That's the implication. But this *experiment* is misleading. What's warming the bottle is the water vapor inside it—it has nothing to do with CO_2."

"That's cheating."

"Yeah, and what Gore is also not teaching is that without the greenhouse effect the planet would be a ball of ice—with a temperature at least 60° colder than today."

"So, the greenhouse effect is a matter of survival."

"Right on, bro'. Now here's Activity Two. 'Carbon dioxide makes up 80 percent of the total greenhouse emissions—' what?" I exclaimed, interrupting myself. "This is corrupt! If you include water vapor," I said, slowing down to make my point, "which makes up ninety five percent of the earth's greenhouse gases, man-made CO_2 accounts for *one hundred sixteen thousandths of a percent* of the greenhouse effect. Take away water vapor, and anthropogenic CO_2 is still only a mere two point three percent of the greenhouse effect!" I was livid. "How does he get away with this?"

Daniel grabbed the poster to read for himself.

"He's got to be referencing this stuff," Daniel said, as he scanned the propaganda. "Unbelievable. The reference is, 'According to *An Inconvenient Truth.*' Gore's *film* is his authority of record."

"That's how they play the game, dude," I shrugged.

"Well, if that's the game Gore's playing, and his followers are citing his movie as the authority, we're all hosed."

BUT NOT IN GREAT BRITAIN

Gore's film is causing global brainwashing. According to a British survey, half of the children in the UK between the ages of seven and eleven are anxious about the effects of global warming and often lose sleep over it.[130] Canada's *Financial Post* has reported: "When Al Gore came to town last week, a mother who had been unable to get tickets called up the University of Toronto and said that her daughter hadn't been able to sleep since viewing *An Inconvenient Truth* at school. She claimed that seeing Mr. Gore in person might make her daughter feel better."[131]

It's happening here, too. Callers to my radio show have said their children are having nightmares after being forced to view Gore's flick in their classrooms.

In England, one brave dad had enough. Stewart Dimmock, a father of two, challenged his government's decision to provide every secondary school in England with a copy of *An Inconvenient Truth* as part of a nationwide environmental campaign. Filing suit in court, his lawyers argued that the movie lacked balance and aimed at influencing, rather than informing, the students.

Amazingly, The United Kingdom's High Court in London agreed with Dimmock and determined that *An Inconvenient Truth* contains "alarmist

---

130 *Community Soundings*, Published by the Primary Review, Faculty of Education, University of Cambridge, 2007.

131 Peter Foster, "Little Miss Apocalypse," *Financial Post*, Wednesday, February 28, 2007.

and exaggerated"[132] content which can only be legally shown to school children if accompanied by a warning regarding the film's blatant "political brainwashing."[133]

In his ruling, Judge Michael Burton said the "apocalyptic vision presented in the film was politically partisan and thus not an impartial scientific analysis of climate change."[134] It was, he continued, the work of a "talented politician and communicator, to make a political statement and to support a political program."[135]

Right on, Judge Burton.

In its synopsis of the movie, the film's distributor, Paramount, makes the judge's case:

Humanity is sitting on a ticking time bomb. If the vast majority of the world's scientists are right, we have just ten years to avert a major catastrophe that could send our entire planet into a tail-spin of epic destruction involving extreme weather, floods, droughts, epidemics and killer heat waves beyond anything we have ever experienced.[136]

Judge Burton diffused the time bomb by mandating the creation of a 56-page instructor's guide, which is now required to be used by teachers in the U.K. in conjunction with classroom viewings of the movie. In its introduction, the guide states:

. . . in parts of the film, Gore presents evidence and arguments which do not accord with mainstream scientific opinion.[137]

132 "British judge says Al Gore's climate change film contains 9 scientific errors," reported by Associated Press, *International Herald Tribune*, October 11, 2007.

133 Sally Peck, "Al Gore's 'Nine Inconvenient Untruths'," United Kingdom Telegraph, October 11, 2007.

134 Ibid.

135 "British judge says Al Gore's climate change film contains 9 scientific errors," reported by Associated Press, *International Herald Tribune*, October 11, 2007.

136 From Paramount Pictures website. http://www.paramountpictures.co.uk/films/an_inconvenient_truth/index.asp.

137 *The climate change film pack — Guidance for teaching staff*, Introduction, published by the United Kingdom Department for Children, Schools and Families, 2007.

Since the vast majority of people exposed to Gore's film worldwide never get to see the British disclaimer, allow me to point out some of the bigger whoppers in Gore's film.

SEA-RISE TERROR

One of the more troubling scenes in *An Inconvenient Truth* involves graphical representations of the world's major cities being submerged due to a rise in sea level from the melting of Antarctica and Greenland. The predictions are pure fiction.

Gore focuses first on Western Antarctica. He calmly claims, "If this were to go, sea levels worldwide would go up 20 feet."

By "go" Gore's implying, *melt*.

He then couples that concocted cataclysm with the total melting of Greenland. Using vague, unreferenced charts supposedly illustrating the amount of melting from Greenland's glaciers and ice sheets, Gore emphatically states, "Because of what is happening in Greenland right now, the map of the world will have to be redrawn." He then prophesies: "If Greenland broke up and melted, or if half of Greenland and half of West Antarctica broke up and melted, this is what would happen to the sea level...."

Playing the role of mad scientist, Gore then graphically reveals Florida flooded, San Francisco swamped, "tens of millions of people" near Beijing displaced, "40 million people" near Shanghai forced to flee and "50 million people" becoming refugees in Calcutta and East Bangladesh.

Adding further to his fright-fest, Gore, in a troubled monotone says, "This is the World Trade Center Memorial Site. After the horrible events of 9/11 we said 'never again.' But this is what would happen to Manhattan." Exquisite Hollywood graphics inundate Ground Zero with a tsunami of water. Gore then claims, "They can measure this *precisely*."

How? He never says. But clearly he seems to be likening global warming to the evils wrought by the Islamo-fascists who flew planes into buildings on 9/11.

Gore tags the scene by asking, "Is it possible that we should prepare against other threats besides terrorists?"

No wonder children are freaking out. Gore is comparing global warming to terrorism.

DAY AFTER TOMORROW

Gore's sea-level terror capitalizes on the 2004 apocalyptic sci-fi film, *The Day After Tomorrow*. Raking in over $543 million at the box office, it's certainly considered a blockbuster, but I'm sure most who saw the film are unable to separate fact from fiction.

The movie opens with NASA's Jack Hall drilling ice core samples on Antarctica's Larsen Ice Shelf. While working, the ice shelf cracks, snapping off from the rest of the continent. Jack survives the climate-induced calamity and, taking advantage of the crisis, scurries to a United Nations global warming conference in New Delhi to present his firsthand account of impending planetary doom. Anthropogenic greenhouse gases are melting the polar icecap, and ocean currents will be altered unless drastic measures are taken—now.

But it's too late. Shortly after the conference, weather buoys in the Atlantic Ocean show a massive drop in water temperature. The melting polar ice has shut off the famed Gulf Stream and North Atlantic current.

Eventually, bizarre circulation changes create an enormous wave, which surges towards Manhattan, submerging the streets under thirty of feet of water. Next, a hurricane of snow slams the city, burying the Statue of Liberty like a wooly mammoth trapped in a giant snowdrift. The new Ice Age has arrived—all because of man's lust for fossil fuels.

Like NASA man Jack Hall, in *An Inconvenient Truth*, Gore warns if the ocean currents were "shut off and the heat transfer stopped," we could head into another ice age. He coyly suggests the melting of Greenland could make that a reality.

Gore is sidestepping the science. Our great oceans will never "shut off" because God has placed *salt* in their waters.

Two of the most important characteristics of ocean water are its temperature and salinity. Together they help govern the density of seawater, which is the major factor controlling the ocean's circulation, both horizontally and vertically. The continual, unstoppable circulation of water occurring below the ocean surface is known as "thermohaline circulation" or THC.

Earth's oceans are immense. They cover 71% of the planet and are incredibly deep. Whereas the average height of the earth's land is 2,755 feet, the average depth of the world's oceans is 12,450 feet, and thanks to the THC that deep pool of water is never stagnant.

Layered with the densest water on bottom and the least dense on top, ocean water tends to move horizontally along lines of equal density. Vertical circulation is primarily limited to the edges of land masses. The general patterns of ocean circulation are akin to massive superhighways of salt water, principally steered by the earth's constant 24-hour rotation; the reliable change of seasons due to our planet's tilted axis in reference to its annual trip around the sun; and the predictable variations in tides caused by the gravitational pull of the moon. Bottom line—ocean circulation will never "shut off."

In the case of the Atlantic and Gulf Stream currents, warm, salty water from tropical latitudes is transported northward, where the naturally colder weather extracts heat from the surface water, thereby allowing it to cool, increase in density, and sink. That water then flows back toward the equator. Similar mechanisms take place across the globe, assuring us that the Statue of Liberty is not going to be suddenly buried in an avalanche of snow.

Occasionally, perturbations in the various currents create alterations that impact weather patterns, but nowhere near the degree that Gore insinuates. One such shift is the periodic El Niño current, which is notorious for causing horrendous flooding on the West Coast. Likewise, its counterpart, La Niña, can cause cold temperature delays in the farming season in the Northern Plains and create conditions that can whip up a 100-year flood in the Missouri and Mississippi Valleys. In addition, as I will explain shortly, a variance known as the Pacific Decadal Oscillation can raise the temperature in the Arctic, melting summer sea ice.

For Gore to even imply that global warming could melt Greenland or West Antarctica, cause a wholesale "shut off" of ocean currents, and swamp millions of people, does not jibe with reality. But since when did reality ever matter to the global whiners?

TRUTH ABOUT SEA LEVEL

Global warming advocates cherish cherry-picking dramatic quotes from the documents procured by the United Nation's Intergovernmental Panel on Climate Change (IPCC). Hundreds of scientists contribute to IPCC reports; some of the scientists are activists, others are quite skeptical about anthropogenic climate change. However, even though the IPCC's editors are pushing the green agenda, there are some unbiased details buried within their papers that defy Gore's scaremongering tactics.

Ever since the end of the last Ice Age, the global sea level has been gradually increasing. The melting ice and snow from that bitterly cold event is continually trickling into our great oceans and seas. According to the often quoted IPCC's Fourth Assessment produced in 2007, over the past 20,000 years sea level has increased nearly 400 feet.[138] While that may sound like a lot, let's break it down further. Over the past century the average sea level rose a mere 1.8 millimeters per year.[139] Try placing your thumb as close to your forefinger as possible, without the two touching—that's the amount of sea level rise each year. Hardly frightening.

Gore's implication that if the West Antarctica ice shelf were to melt, sea level would increase 20 feet is obviously cherry-picked out of context from the IPCC, which, as part of an obvious set of improbable, hypothetical scenarios states, "the collapse of [the West Antarctica Ice Sheet] . . . would trigger another five to six metres [16–19 feet] of sea level rise . . . that would take many hundreds of years to complete."[140]

What Al Gore failed to mention was the IPCC section from which his West Antarctica and Greenland melting scenarios were lifted is entitled, *How Likely are Major or Abrupt Climate Changes, such as Loss of Ice Sheets of Changes in Global Ocean Circulation?* The section begins with this surprising disclaimer:

---

138 International Panel on Climate Change, Fourth Assessment Report (2007), Chapter 5, p. 409.
139 Ibid., p. 821.
140 Ibid., p. 819.

Abrupt climate changes, such as the collapse of the West Antarctic Ice Sheet, the rapid loss of the Greenland Ice Sheet or large scale changes of ocean circulation systems are not considered likely to occur in the 21st century....[141]

That means that over the next 90-years, even the usually enthusiastic global warming editors at the U.N. are not anticipating the kind of predictions that Gore lays forth in his film.

The IPCC also states, "All studies for the 21st century project that antarctic[142] [ice] changes will contribute *negatively* to sea level"[143] (italics mine). That means despite the dramatic clips Al Gore shows of glaciers on the Antarctic Peninsula calving and crashing into the sea, there is more snow and ice accumulating on Antarctica than is breaking off and melting into the surrounding waters. In addition, their report declares, "... for the last two decades ... Antarctica as a whole has not warmed."[144]

Gore's fancy photos from the Antarctic Peninsula (which comprises less than 2% of the continent, and extends into the warmer waters of the South Sea) don't reveal that the *cooling* of the Antarctic has posed quite a conundrum to the global warmers for some time. In a leaked CRU email sent to NASA's Jim Hansen in 1999 we read:

We are trying to understand your data relative to Phil Jones's in the Southern Hemisphere during the 1990s ... I have been perusing your website and have noted the most recent years are cold over Antarctica in your dataset. This could be the focus of the problem....

If Antarctica cools, there will be consequences for the Southern Hemisphere atmospheric circulation patterns, conceivably even contributing to the recent cooling of marine air temperature relative to sea surface temperature.

Yeah, and if word gets out that Antarctica is cooling, there could be

141 Ibid., p. 818.
142 The IPCC report spelled "antarctic" in lower case.
143 International Panel on Climate Change, Fourth Assessment Report. Chapter 10, 2007, p. 816.
144 Ibid.

consequences for your buddy Al Gore's net worth, as well as your own, Mr. Hansen.

Regarding the potential collapse of the West Antarctic Ice Sheet, even the IPCC admits, ". . . no quantitative information is available from the current generation of ice sheet models as to the likelihood of timing of such an event."[145] And, regarding the swamping of major cities projected by Gore, the report reads, ". . . accelerated sea level rise caused by rapid dynamic response of the ice sheets to climate change is very unlikely during the 21st century."[146]

As for the liquidation of Greenland, the IPCC states: ". . . the total melting of the Greenland Ice Sheet, which would raise global sea level by about seven metres, is a slow process that would take many hundreds of years to complete . . . no quantitative information is available from the current generation of ice sheet models as to the likelihood or timing of such an event."[147]

Translation: the total melting of Greenland is a pipedream, and Gore's Antarctic machinations are madness.

SINKING, OR STINKING ISLANDS?

For decades there have been ridiculous rumors of islands sinking due to rising waters. Gore's film seems to show images of such islands and claims that because of melting polar ice, "the citizens of these Pacific nations all had to evacuate to New Zealand."

Gore's assertion was a patent lie—a lie that first started floating the boats of eco-freaks back in the Eighties.

In October 1987, Maumoon Abdul Gayoom, the Muslim dictator of the south Asia island chain, the Maldives, presented an impassioned address to the United Nations General Assembly, declaring his Indian Ocean nation of some 250,000 citizens was threatened by a rising sea. He claimed that "a mean sea level rise of two meters would suffice to

---

145 Ibid., p. 819.
146 Ibid., p. 817.
147 Ibid., p. 819.

virtually submerge the entire country of 1,190 small islands, most of which barely rise over two meters above mean sea level. That would be the death of a nation."[148] Mr. Gayoom insisted his dire status was related to climate changes that had been "provoked and aggravated by man."

Environmentalists lapped it up.

Furthering the hype, in 2001 the leaders of Tuvalu—midway between Hawaii and Australia—announced they wanted to evacuate because of the global warming/rising ocean threat. After being rebuffed by Australia, the Tuvaluans asked New Zealand to accept its 11,000 citizens, but New Zealand also declined.

I guess, come editing time, Al never got the news.

Truth is, Tuvalu and Maldives are *not* being swamped by an unstoppable rising sea. If anything, in that part of the world it appears that the ocean levels are *falling*.

In 2004 Stockholm University professor Nils-Axel Mörner of Sweden published a paper in *Global and Planetary Change* (hardly a bastion for global warming deniers) regarding his extensive research of the ocean around the Maldives. He noted, "In our study of the coastal dynamics and the geomorphology of the shores we were unable to detect any traces of a recent sea level rise. On the contrary, we found quite clear morphological indications of a recent fall in sea level."[149]

His research indicated that sea level about the Maldives has fallen approximately 11 inches in the past 50 years. In fact, additional research indicates that about the time the leaders of Tuvalu created headlines in 2001, the sea level surrounding the nine atoll islands of their country had recently *fallen* 2.5 inches.[150]

A study by honest experts at Tuvalu's Meteorological Service also reported similar sea level falls had been recorded in the nearby Nauru and the Solomon Islands.[151]

---

148 Address by Mr. Maumoon Adbul Gayoom, before the 42nd session of the United Nations General Assembly, October 19, 1987.

149 Nils-Axel Mörner, Michael Tooley, and Göran Possnert, "New perspectives for the future of the Maldives," *Global and Planetary Change* Vol. 40, Issues 1–2 (January 2004): pp. 177–182.

150 Mark Chipperfield in Tuvalu and David Harrison in London, "Falling Sea Level Upsets Theory of Global Warming," London *Telegraph*, August 6, 2000.

151 Ibid.

Professor Patrick Nunn, who researches sea level rise at the University of the South Pacific in Fiji, admitted, "A lot of these sea gauges have been slowly falling over the last five years. . . ."[152]

So, why the hue and cry about these islands sinking? Perhaps because of dumb development and living in a relative pig sty.

The Maldives are relatively flat atolls, composed of coral. Tourism was only introduced to Maldives in 1972 with the opening of the plush Kurumba Village Resort on the North Malé Atoll. Now there are 87 resorts jammed primarily atop three islands: the North and South Malé Atolls and the Ari Atoll. Tourism has become the Maldives' leading industry, and locally mined coral rock has been the main aggregate for constructing these resorts.[153] Digging up coral to build large hotels and conference centers is as stupid as sucking the air out of a life raft to breathe. The mining has severely compromised the atolls, creating the impression that the islands are sinking.

Likewise, Tuvalu's problem is not climate change, nor tourism—they only receive about 1000 tourists each year. Tuvalu's mess is that their country was never meant for modern habitation. Their primary indigenous vegetable crop, taro, has been gravely overfarmed. There is no fresh water available—only what can be cached from rain. Much of the population on the main island uses a lagoon for its bathing and toilet facilities. The tiny country ships its commercial waste to landfills in Fiji and New Zealand. Tuvalu is a tropical island mess being run by imbeciles who are using global warming as a shakedown operation, the likes of which would make a Chicago community organizer proud. In a 2007 press release, the Tuvalu government said:

> The Deputy Prime Minister of Tuvalu, the Honorable Tavau Teii, said that major greenhouse polluters should pay Tuvalu for the impacts of climate change. This claim was made during his speech to the United

---

152 Ibid.

153 Abdulla Naseer, "Status of Coral Mining in the Maldives: Impacts and Management Options," Marine Research Section, Ministry of Fisheries and Agriculture Malé, Republic of Maldives. Paper located on Food and Agriculture Organization of the United Nations website, June 16, 2009. http://www.fao.org/docrep/X5623E/x5623e0o.htm.

Nations High Level Meeting on Climate Change held at the U.N. head-quarters in New York.

"Tuvalu is highly vulnerable to the impacts of climate change so we are seeking new funding arrangements to protect us from the impacts of climate change," Mr. Teii said. "Rather than relying on aid money we believe that the major greenhouse polluters should pay for the impacts they are causing."[154]

Caving to public pressure, in April of 2009 (three years after the release of Gore's film), New Zealand finally responded to the phony cries of goo-goo activists who claimed if nothing was done the Tuvaluans would soon be blubbing with the fish, allowing them to immigrate. I'm confident Gore is now citing the evacuation as a prophecy fulfilled.

ARCTIC ICE SHRINKAGE?

In Gore's film, he refers to the Arctic as a "canary in the coal mine . . . I say canary in the coal mine," he spouts to the camera, "because the Arctic is one part of the world that is experiencing faster impact from global warming." His proof includes a submarine trip he claims to have taken, to the North Pole:

I went up to the North Pole. I went under that ice cap in a nuclear sub-marine that surfaced through the ice like this. This thing started patrolling in 1957. They have gone under the ice and measured with their radar look-ing upward to measure how thick it is because they can only surface where the thickness of the ice is 3 and half feet thick or less. So they have kept a meticulous record and they wouldn't release because it was national secu-rity. I went up there in order to persuade them to release them, and they did. And here's what that record showed: Starting in 1970, there was a pre-cipitous drop off in the amount and extent and thickness of the arctic ice cap. It has diminished by 40% in 40 years. There are two studies showing that in the next 50 or 70 years in summertime it will be completely gone.

---

154 Press Release, "Tuvalu calls for Climate Change Polluters to Pay," Nation of Tuvalu, September 29, 2007.

While it's difficult to confirm Gore's personal submarine trip and the navy supposedly bowing to his demands to release classified information for use in a Hollywood film, we can confirm that his canary analogy is absurd. First of all, why base history on submarine observations that only stretch back to 1957? Arctic temperature records indicate that the warmth of the 1930s equaled or exceeded average temperatures recorded there after 1970.[155] As for the ice, no one officially began measuring it until satellites became available in 1979. However, since Gore delights in the anecdotal, it's interesting that he passed on using this navy photograph from 1959, showing the U.S. sub "Skate," floating like a cork in iceless water at the North Pole (see Figure 6.1).

Figure 6.1: U.S.S. *Skate* (SSN-578), surfaced at the North Pole, March 17, 1959.

Source: NAVSOURCE

In a comment originally posted by the profound global warming blogger John Daly, a former U.S.S. *Skate* crewmember said that in 1959 "the *Skate* found open water both in the summer and following winter. We surfaced near the North Pole in the winter through thin ice less than two feet thick."[156]

Daly, by the way, was the brilliant blogger whose death was regarded as "cheering news," in a CRU email written by Phil Jones.

Since sea ice has only been monitored via satellite since 1979, it is important to note that when you hear news reports of "the least amount of ice in history," the history they speak of is just slightly over 30 years— they're not including the older personal accounts by former Navy men.

155 Roy W. Spencer, *Climate Confusion* (New York; Encounter Books, 2008), p. 82.
156 James F. Hester, personal email communication, December 2000, from "Top of the World: is the North Pole Turning to Water," John Daly, http://www.john-daly.com/polar/arctic.htm.

In 2007, you probably heard about the most expansive Arctic ice melt ever, but were you told of the record refreeze that autumn? During a ten-day period in November, a NASA eye-in-the-sky recorded sea ice in the Arctic Ocean *growing 58,000 square miles per day*—about the same size as Illinois or Georgia.[157]

In the summer of 2008 and 2009, the Arctic ice melts were less significant, but the fall refreezes were nearly as dramatic as that witnessed in 2007.

Undaunted, since receiving his Oscar, Gore has now speeded up his predictions of an ice-free Arctic. While lecturing a German audience, he claimed that "the entire north polar ice cap will be completely gone in five years."[158]

Another fact gone unreported is that the geographical boundaries of the Arctic have been expanded, thus increasing its stated average temperature. Previously, "Arctic" referred to the Arctic Circle, roughly 66 degrees north latitude. This region of the globe includes portions of Alaska, Canada, Greenland, Sweden, Norway, Finland, Russia, and Iceland. The fringe of Arctic Circle experiences 24 full hours of sunlight on the June 21st summer solstice, and 24 hours of darkness during the winter solstice December 21st.

However, several years ago the Arctic Climate Impact Assessment (ACIA), funded by the United States and working in lockstep with the United Nation's Intergovernmental Panel on Climate Change, decided to expand the Arctic by about 50%, or approximately 4 million square miles. By artificially expanding the Arctic beyond 66 degrees north latitude, naturally warmer temperature monitoring stations are now included in determining the average temperature of the region. Besides making it easier to hype a supposed shift in climate, environmentalists are also using the newly defined Arctic region as a tool to prevent oil exploration and protect the supposedly endangered polar bear.

---

157 "Arctic Sea Ice Re-Freezing at Record Pace," *Associated Press,* December 12, 2007.
158 "Al Gore: The North Polar Ice Cap Will Disappear in Five Years," YouTube video, posted December 13, 2008, http://www.youtube.com/watch?v=KrPCUWWjh0c.

PHONY POLAR BEARS

Another fib from Gore's film is that "a new study shows that for the first time they're finding polar bears that have actually drowned, swimming long distances of up to 60 miles to find the [Arctic] ice."

His lines are accompanied by cheesy animation of what appears to be a narcoleptic polar bear struggling to climb aboard a lonely sliver of sea ice. The weight of the bear snaps the thin ice chip like a Rye Crisp. The polar bear is doomed to drown. This too, is another fascinating work of fiction.

It has always been assumed that Gore was referring to the 2004 account of the four polar bear carcasses that were found floating in the Beaufort Sea off the coast of Alaska. The dead animals were spotted by researchers from the U.S. Minerals Management Service (USMMS), who, for the past 25 years, have regularly surveyed the polar bear population of that region. Because the animals are instinctively hearty swimmers, the USMMS says that, prior to 2004, drowned polar bears had never been documented in their surveys. However, in late September of that year, a USMMS helicopter crew witnessed four carcasses, presumed to have somehow succumbed in the sea. Word immediately got out to activists, who quickly blamed it all on anthropogenic global warming.

Truth is, according to the National Weather Service in Fairbanks, Alaska, at the time of the observation, a normal amount of sea ice was covering the Beaufort.[159] Therefore, it was unlikely that the drowned bears ran out of ice to rest upon. So, what happened? I personally suspected a fierce storm and dug into records of the National Weather Service. Sure enough, according to a September Weather Service summary, there was a potent, three-day storm that whipped up some pretty wild waves.

The most significant event of the month was the combination of a low pressure area that moved from Bristol Bay east over the Gulf of Alaska on the 19th and 20th, and into the Canadian Great Lakes on the 21st, and a western Aleutian low that moved over the central Bering Sea on the 19th and 20th, and into southwest Alaska on the 21st. These developments

---

159 *A Look Across the State*, Fairbanks Forecast Office National Weather Service, September, 2004 Summary, http://climate.gi.alaska.edu/Statewide/2004/Sep04.html.

brought vigorous westerlies and rough seas to the Aleutians and the Gulf of Alaska.[160]

The polar bears were not killed by a lack of sea-ice rest stops; most likely they were swamped by three days of high winds and rough seas. Gore conveniently left out that part of the story.

TRUTH ABOUT KILIMANJARO

If you ever want to pull heart strings, just show photographs of impoverished children in Africa. Most of us will write a check to whatever the cause in a heartbeat. Gore pulls similar strings with his claim that two of Africa's icons, Mount Kilimanjaro and Lake Chad, are doomed because of global warming.

Comparing photographs of the 19,340-foot Mount Kilimanjaro taken in 1928 and 2000, Gore deceptively illustrates how the glacier atop the former volcano has noticeably decreased. At an unsuspecting glance, the decrease is stunning. However, Al conveniently fails to tell his audience when most of the ice actually melted.

In November 2007, Britain's *Guardian* ran a story that boldly began, "A new study on the dwindling ice cap of Africa's highest peak, Mount Kilimanjaro, suggests that global warming has nothing to do with the alarming loss of its beautiful snows."[161]

The study, published in the *American Scientist*, verifies local legend that the melting of Kilimanjaro was noticed a few decades after the conclusion of the Little Ice Age. The research also stated that, while Kilimanjaro's glacier had declined 90% between 1880 and 2003, "the decline in Kilimanjaro's ice has been going on for more than a century and that most of it occurred before 1953."[162]

Furthermore, to blame the shrinkage on warmer temperature is

---

160 Ibid.
161 Gadiosa Lamtey, "Mt. Kilimanjaro ice still baffles experts," *Guardian*, November 26, 2007.
162 Ibid.

shortsighted. According to the study, "Another important observation is that the air temperatures measured at the altitude of the glaciers and ice cap on Kilimanjaro are almost always substantially below freezing (rarely above −3°F). Thus, the air by itself cannot warm ice to melting. . . . When pieced together, these disparate lines of evidence do not suggest that any warming at Kilimanjaro's summit has been large enough to explain the disappearance of most of its ice, either during the whole twentieth century or during the best-measured period, the last 25 years."

In his film, Gore refers to his friend Lonnie Thompson, a climatologist from Ohio State University. Gore implies that Thompson has told him the snow-topped peaks of Kilimanjaro will vanish "within the decade." Typical of Gore's penchant for playing with predictions, I discovered an interview conducted by the German magazine, *Der Spiegel* in 2006, in which Thompson gives the glaciers twenty years.[163]

For Gore, *10* obviously sounds more threatening than 20, but he also leaves out some other info from his buddy. Thompson was the first scientist to effectively drill the ice cores on Kilimanjaro. His research found that the African glaciers first developed 11,700 years ago;[164] about the same time the last Ice Age ended—in other words, they've been melting ever since. In fact, in *Spiegel's* interview with Thompson, the magazine reported that Thompson believes, "more than 80% of the tropical glaciers, including those on Mt. Kilimanjaro . . . have already vanished within the last hundred years."[165]

Certainly that can't be blamed on CO_2 emissions, but it can be explained by natural, gradual, post-Ice Age warming, and perhaps another variable.

Climatologist Doug Hardy, a colleague of Thompson, says, "The phrase global warming is misleading, as are the alarmist reports of the complete disappearance of all glaciers on Kilimanjaro."[166]

Hardy points to Kilimanjaro's lack of snowfall.

"Dryness, not warming, is what's causing the glaciers to recede," says Hardy.[167] Indeed, precipitation in Africa has been known to vary greatly

---

163 Hilmar Schmundt, "Why Is Mt. Kilimanjaro Melting?" *Der Spiegel*, February 20, 2006.
164 Ibid.
165 Ibid.
166 Ibid.
167 Ibid.

throughout history, with droughts often lasting hundreds of years. Over the last century most agree precipitation has decreased 33%.

In addition, Kilimanjaro's problems may also be exacerbated by the disregard locals have for their iconic mountain.

German biologist Andreas Hemp, who studies the jungles at the base of Kilimanjaro, notes that corresponding to the decrease in rain and snow over the past century, "the number of people living at the base of the mountain has grown twenty-fold, or to about a million. The forest suffers as a result," says Hemp. "Illegal loggers are assaulting the rain forest from below, and fires have lowered the upper range of the evergreen forest by about 500 meters in the last 30 years."[168]

It is quite likely that such radical changes in vegetation at the base of the giant mountain have altered the micro-climates, shifting winds, and thus contributing to the lack of annual precipitation. In fact, according to another study in *American Scientist*, "It is certainly possible that the ice cap [atop Kilimanjaro] has come and gone many times over hundreds of thousands of years."[169]

However, even when precipitation patterns eventually shift and the snows of Kilimanjaro fall in abundance, the media will blame that on global warming, too. Hence, the recent BBC headline: "Global Warming Boost to Glaciers."[170]

LAKE CHAD SLEIGHT OF HAND

In a further attempt to play on African pity, Gore says of the continent,

Unbelievable tragedies have been unfolding there, and there are a lot reasons for it. Darfur and Niger are among those tragedies. One of the factors that have been compounding this is the lack of rainfall and the increasing drought. This is Lake Chad, once one of the largest lakes in the world. It has

---

168 Ibid.
169 Philip W. Mote, Georg Kaser, "The Shrinking Glaciers of Kilimanjaro: Can Global Warming Be Blamed?" *American Scientist* (July-August 2007).
170 *BBC News*, August 24, 2006.

dried up over the last few decades to almost nothing. That has been complicating the other problems that they also have.

The horror in Darfur and Niger have to do with conniving politicians, deadly dictators, ethnocentric tribal leaders, and radical Muslims—*not* global warming. But by this time Gore has his audience in a trance. With slick Hollywood sleight of hand, he quickly shows another deceitful series of photographs which, just like with Kilimanjaro, make it appear Lake Chad is vanishing before their eyes.

What Gore conveniently removes from the record is the fact that Lake Chad is really not a lake at all; it's a giant puddle that geologists estimate has completely dried out at least six times over the past 1,000 years.[171] Even during wet years, Chad is only about 23 feet deep.

Ninety percent of Lake Chad's water is fed by the Chari River, which swells seasonally by runoff from the meager rains that typically fall from April to September.[172] Because of its shallowness, Chad is subject to exaggerated fluctuations in expanse, depending on the rains. The current shrinkage, which has been underway since the 1960s, can be blamed only in part because of drier weather. The biggest problem is that more water is being pulled from the river to meet the needs of a massive population boom.

In a 2001 NASA press release (that Gore ignored) the agency explained:

> Over the last 40 years, the discharge from the Chari/Logone river system at the city of N'Djamena in Chad has decreased by almost 75%, drastically reducing the input into the lake. Between the increase in agricultural water use and the drier climate, there has been a massive decline in the amount of water in Lake Chad.[173]

The agricultural demands on the river have been driven by a population that has increased from just over 100,000 in 1960 to approaching 750,000 today. The ballooning population alone is sucking Chad dry. However, as

---

171 Japan Aerospace Exploration Agency, Internet Press Release, January 17, 2005. http://www.eorc.jaxa.jp/en/imgdata/topics/2005/tp050117.html.
172 The annual rainfall in the Chari basin is 20 inches.
173 Goddard Space Flight Center, Press Release Number 01-17, February 17, 2001.

history has proven, there will be especially wet years ahead for the Chari River basin and Lake Chad will re-expand. But will Gore show us *those* photographs when that happens? Don't hold your breath.

AL'S HOT AIR

Scenes from Hurricane Katrina, the 2005 monster storm that swamped Louisiana and Mississippi, are also prominently featured in Gore's propaganda piece, complete with sirens blaring, residents screaming, and New Orleans Mayor Ray "get off your asses and let's do something" Nagin losing his cool.[174]

Gore cleverly suggests that the cause of Katrina was an ocean heated by global warming: "When the oceans get warmer, that causes stronger storms." He then asks, "How in God's name could this happen here? There had been warnings that hurricanes would get stronger."

We'll discuss hurricanes in more detail in the following chapter, but first, allow me to say that one of God's great gifts to the earth is the hurricane. Among other benefits, hurricanes are an efficient conduit for hot equatorial air to be transported to colder parts of the planet, thus balancing out earth's average temperature. However, to claim that warmer water creates stronger storms is wholly misleading. Yes, hurricanes require warm water of at least 82 degrees in order to form, but once the water gets much above 82, there is no correlation to the formation of stronger hurricanes.

A gentleman I have interviewed many times, Dr. William Gray, a pioneer in hurricane research and for decades the foremost hurricane forecaster, has stated, "One of the most misunderstood topics in tropical meteorology is association of Sea Surface Temperatures and tropical cyclone [a.k.a. hurricane] frequency and intensity . . . changes in Sea Surface Temperatures explain just a small portion (about 10%) of the variance of seasonal and monthly hurricane activity."[175]

---

174 Mayor Ray Nagan made the "get off your asses and let's do something" comment during an interview on New Orleans radio station WWL, and conducted by Garland Robinette, September 1, 2005.

175 Dr. William Gray, "Global Warming and Hurricanes," Colorado State University, April 5, 2007.

Gore even tries to claim that a freak South Atlantic Ocean storm that struck Brazil in March 2004, was a hurricane that developed in direct response to global warming—even though the meteorological community questions whether the tempest was, by definition, a hurricane. "The science textbooks," he laments, "have to be rewritten because they say it is impossible to have a hurricane in the South Atlantic."

What deceptive fraud.

The storm in question was named Catarina, because it came ashore in the Brazilian state of Santa Catarina, and it's been the topic of much debate. Catarina's story begins in January and February of 2004, Brazil's coldest months in 25 years. According to University of Sao Paolo scientist Pedro Leite da Silva, sea surface temperatures were also slightly cooler than average going into mid-March.[176] Initially Catarina was a nameless low pressure system (like the one your TV meteorologist often talks about) well off the Brazilian coast. Before long it drifted into an unusual zone between the subtropical and polar jet streams. Once there, the storm slowly took on tropical hurricane characteristics and raced inland to the Brazilian state of Santa Catarina on March 28, with winds blowing 85 miles per hour.

Professor Lance Bosart of the Albany State University, who has spent considerable time researching Catarina, believes the storm wasn't a unique occurrence at all, although its landfall was certainly historic. "You have to be extremely arrogant to assume a storm like Catarina hadn't happened in the South Atlantic before the satellite era."[177]

Extremely arrogant assumptions? Sounds like shades of "Little Al." It also sounds like science textbooks won't be rewritten anytime soon.

CRISPY CORAL

Gore also spends time talking about crispy coral: "Coral reefs all over the world, because of global warming and other factors, are bleaching, and

---

176 Bob Hudson, "What was Catarina?" *UCAR Quarterly*, Summer 2005, University Corporation for Atmospheric Research, National Center for Atmospheric Research, Boulder, Colorado.
177 Ibid.

they end up like this." The accompanying film shows a decaying chunk of coral. Again, the viewer is being deceived into thinking that global warming is killing these wonderful living organisms. It's not so.

Coral is an abundant life form that creates ideal habitat for countless sea creatures. In most ocean settings coral thrives, even growing plentifully upon junked ships and oil platforms. Like all living organisms, coral eventually dies. Death can occur from exposure to air, sunlight, disease, breakage, or abnormal conditions—like those that occurred in 1998.

During the last big El Niño in the late Nineties, the waters of the Eastern Pacific warmed several degrees above average. I recall fishermen off the coast of San Francisco reporting spectacular schools of marlin, with their giant spiny dorsal fins and spearlike snouts, zipping through the waters off the Golden Gate Bridge—a spectacle usually only seen 400 miles south, near Los Angeles. During this rare episode of El Niño warmth, some Pacific coral "bleached," (i.e., fried and died). However, since the Nineties, coral bleached by warmed water is tough to find.

Further demonizing carbon dioxide and trying to pin coral destruction on it, Gore warns a doubling of CO_2 levels will acidify seawater, wiping out all the world's coral reefs by 2050. This, too, is not plausible. The scientific consensus is clear: the first known coral reefs have been found in Mesozoic sediment, when atmospheric CO_2 levels averaged rates three to six times greater than today.

But any honest consensus that defies Al Gore's is deemed a minority position.

7

MYTH OF CONSENSUS

The only thing more dangerous than ignorance is arrogance.

—Albert Einstein

"**RON IN SAN JOSE**. Thanks for calling. You're on the air."

"Yeah, Sussman, I realize you used to be a TV weatherman, and you talk a lot about science and global warming, but I'd like to see you find more than five *real* scientists who believe there's no global warming."

"Ron, I'm going to let you do some homework—"

"Five, you couldn't find more than five—"

"Ron, please, calm down and just listen—you might learn something here. I know you're calling on a car phone, so when you get home I want you to do an Internet search for The Heidelberg Appeal. Are you with me, Ron?"

"There are no more than five—"

"Ron, pipe down and listen up. When you get home look up 'Heidelberg Appeal.' H-E-I-D-E-L-B-E-R-G Appeal. When you do, you'll see it was a response to the ridiculous 1992 Rio Earth Summit, where global warming was first made a political issue. All these years later, it's now been signed by 4,000 scientists, including 72 Nobel Prize winners—all skeptics of man-caused global warming. Did you hear me, Ron?"

"You couldn't find more than five—"

"Ron, the Heidelberg Appeal: 4,000 scientists, 72 Nobel Prize winners, signing a declaration which basically states, 'let's stick to science and not *ad hominem* poppycock.' How's that grab you, Ron? Ron?"

BEEEEEEEP.

Ron hung up. He couldn't handle my challenge—callers like Ron never can. Despite the facts proving otherwise, they want to believe that there is a consensus among scientists and weather professionals that anthropogenic global warming threatens the earth and that only a radical reduction in fossil fuel consumption will save it. There is no such consensus.

The 1992 Heidelberg Appeal was the first petition of its kind, and one that the mainstream media purposefully ignored. Instead, we are fed a regular diet of unsupportive statements from political puppets who claim the Emperor has a splendid suit of clothes.

Look no further than the global warming front persons from my own whacky state. Senator Diane Feinstein, one of the Senate's most influential members, claims the debate on global warming has "shifted considerably because people know it's real. The science has coalesced."[178]

Senator Barbara Boxer, who chairs the powerful Senate Environment and Public Works Committee, asserts she has known for some time that with "the overwhelming science out there, the deniers' days were numbered."[179]

Even the former great bodybuilder-turned-actor, and now futile governor, Arnold Schwarzenegger, conclusively states: "I say the debate is over. We know the science, we see the threat, and we know the time for action is now."[180]

But despite the Terminator poking his beefy finger into the public's chest and declaring the debate is over; Boxer using her bully pulpit to name call and marginalize; and the egalitarian Feinstein implying that scientists have come together like beads of mercury in a thermometer to form unanimous agreement that the earth is boiling over, *there is no consensus.*

---

178 Juliet Eilperin, "Lawmakers on Hill Seek Consensus on Warming," *Washington Post*, January 31, 2007.
179 Sharon Begley, "The Truth About Denial," *Newsweek*, August 15, 2007.
180 Remarks by Arnold Schwarzenegger, United Nations World Environment Day Conference, San Francisco, June 1, 2005.

There is an even more dramatic statement refuting global warming that I could have shared with Ron the caller: the Oregon Petition.

Created just after the United States signed the United Nation's Kyoto Global Warming Treaty in 1997 (while the Earth Summit in Rio addressed global warming as fact, Kyoto gave teeth to the issue by attempting to force the United States into reducing carbon dioxide emissions), the Oregon Institute of Science and Medicine comprised their petition in two succinctly prepared paragraphs:

> We urge the United States government to reject the global warming agreement that was written in Kyoto, Japan in December 1997, and any other similar proposals. The proposed limits on greenhouse gases would harm the environment, hinder the advance of science and technology, and damage the health and welfare of mankind.
>
> There is no convincing scientific evidence that human release of carbon dioxide, methane, or other greenhouse gasses is causing or will, in the foreseeable future, cause catastrophic heating of the Earth's atmosphere and disruption of the Earth's climate. Moreover, there is substantial scientific evidence that increases in atmospheric carbon dioxide produce many beneficial effects upon the natural plant and animal environments of the Earth.

The Oregon Petition was eventually signed by over *30,000* American scientists who "have formal training in the analysis of information in physical science."[181]

Despite what the politicians, activists, and their bullhorns in the media tell you—there is *no* consensus. In fact, science, by consensus, is brain-dead.

ORIGINAL DENIERS

As presented in Chapter 2, in order for the science to work properly, someone must dare question the hypothesis at hand. However, in the same way that happens when one dares to oppose the regime of a demagogue,

---

181 Oregon Petition signatory instructions, Oregon Institute of Science and Medicine, http://www.oism.org/pproject/GWPetition.pdf.

questioning the global warming hypothesis produces immediate back-lash, such as research funding drying up or academic tenure denied. Sadly, I know many TV weather-folks who would never reveal their denier credentials out of fear of being passed over when it comes time for promotion.

Those who persecute the deniers and skeptics of anthropogenic climate change are playing a fiendish game of retribution that places them amongst some cagey characters.

Around 150 A.D., Claudius Ptolemy, an Egyptian living in Alexandria, gathered and organized the thoughts of earlier scientific thinkers and published his theory of the universe. Ptolemy believed the earth was a fixed, immovable mass, located at the very center of the universe, and all celestial bodies, including the sun and the stars, revolved around it. Like global warming, it was an arrogant theory that appealed to the human ego; Ptolemy believed the universe revolved around *man*. In the same way, the global whiners swear *man* has the power to not only alter the climate, but also the knowledge to fix it.

Since the telescope would not be invented for another 1,500 years, it was difficult in Ptolemy's time to conclusively dismantle his theory. The theory eventually became known as Ptolemy's System and soon became regarded as scientific law to be taught in all institutions of advanced learning. Questioning Ptolemy's view of the universe carried severe consequences, immediately branding all such skeptics as heretics. Punishment included excommunication from the Church, banishment from one's country, and, depending on the heretic's temperament, the ruling ecclesiastic authorities could prescribe a public burning at the stake. Because of the obviously intense pressure to conform to Ptolemy's worldview adopted by the ruling elite, absolutely no debate was tolerated, thus, giving an *appearance* of consensus; deniers were forced to secretly discuss their own contradictory theories of the universe in the shadows.

Finally, 1,400 years later, there was a major breakthrough. Polish mathematician and church official, Nicolas Copernicus, spent decades privately defying the centuries-old cover-up. Through extensive research, much of which was conducted under the cover of night atop the roof of his church, without the aid of a telescope (the invention of which was still 60 years away), Copernicus asserted that the earth rotated on its axis once *daily* and traveled around the sun once yearly, and was *not* the hub of the universe. Of course, he was spot on.

Copernicus' findings were quietly published in his brilliant 1530 research manuscript, *De Revolutionibus Orbium Coelestium*, which he privately shared with trusted colleagues who were intrigued and challenged by his detailed work. Fortunately, Copernicus died of natural causes in 1543. Had he lived another decade or two, his last breath might well have been taken from him by the flames of the stake.

Following his passing, Copernicus' revelatory work, *De Revolutionibus,* began to have an impact on the thinking of many subsequent great minds. Italian scientist Giordano Bruno was convinced of the Copernicus system and, like an inspired evangelist, dared to openly share the theory. Because of his zeal for the truth, Bruno was arrested by the authorities in 1592 and placed on trial. Refusing to renounce his beliefs in Copernicus' principles, Bruno was burned at the stake, and *De Revolutionibus* was placed on the List of Forbidden Books.

Galileo was another Ptolemy denier. Utilizing observations gathered with his new invention, the telescope, in 1632, he boldly published his *Dialogue Concerning the Two Chief World Systems*, a work that upheld the Copernican system over Ptolemy's. Galileo was brought before the religious inquisitors in Rome and commanded to renounce all his beliefs and writings supporting Copernicus' theory, but it was too late. Though very primitive, Galileo's telescopes were being mass produced, and too many eyeballs were confirming the truth that our planet and solar system were plainly rotating about the sun. The mandated consensus, which had held fast for nearly 1,500 years, was finally broken.

Thank God for brave souls like Copernicus, Bruno, and Galileo. Their lives provide a fount of inspiration today for those who find themselves defying a cultlike suppression of the truth.

MEDIA BLACKLIST

"Dude, what was that all about?" asked my trusted meteorological assistant.

"I don't know—but I feel like I've been mugged," I replied, sinking into the padded leather chair behind my desk in the Channel Five weather office.

It was 1996. I had just gotten off the phone with an odious guy claiming to be a reporter with a Florida newspaper. He said he was writing a story about the Leipzig Declaration on Global Climate Change, of which I had recently become a signatory. Leipzig challenged the soon-to-be announced Kyoto Treaty on global warming, and strongly condemned the 1992 Treaty presented in Rio. The scientific reasoning behind both of these treaties was, in my opinion, junk—and I was obviously not alone in that assessment.

The heart of the Leipzig Declaration stated:

> We consider the scientific basis of the 1992 Global Climate Treaty to be flawed and its goal to be unrealistic. The policies to implement the Treaty are, as of now, based solely on unproven scientific theories, imperfect computer models, and the unsupported assumption that catastrophic global warming follows from an increase in greenhouse gases, requiring immediate action. We do not agree. We believe that the dire predictions of a future warming have not been validated by the historic climate record, which appears to be dominated by natural fluctuations, showing both warming and cooling. These predictions are based on nothing more than theoretical models and cannot be relied on to construct far-reaching policies.

The supposed journalist speaking with me on the phone initially began his conversation in a friendly sort of way, but his tone quickly changed. What I thought was going to be a professional interview regarding the content of Leipzig and my support for this thoughtful declaration, quickly transformed into a fiery inquisition. The signers of Leipzig were "rebels," he roared.

At the time I was certainly a skeptic of man-caused global warming—I am not sure if I would have considered myself a total denier yet, but big-time skeptic, yes. I was becoming interested in the more recent temperature record, which revealed the warmest decade was the 1930s, when industrial-based CO_2 was *lower* than it is today, and the global cool-down between 1940 and 1970, which coincided with a slight *increase* in CO_2. For me, linking climate changes with the burning of fossil fuels seemed rather presumptuous, because it seemed like a simple matter of sloppy and dishonest science.

"So, you don't believe humans are influencing climate?"

"I don't think we can be sure of it," I calmly replied. "The science isn't there."

"And you don't believe there's global warming?"

"The jury is still out," I said.

The reporter then launched into a tirade about there being a "consensus among real scientists" which, he inferred, did not include mere television meteorologists like me. Somehow he thought my TV status should disqualify me from signing Leipzig and prevent me from having a public opinion on global warming.

"What kind of crappy excuse for journalism was that?" I wondered aloud to my assistant as I hung up the phone.

Leipzig was a bold declaration, and I was proud to have been personally asked by its author, the man known to Al Gore as the "Chief Denier," Dr. Fred Singer, to sign on. Even though Kyoto was signed by the Clinton-Gore administration, thanks to the Heidelberg, the Leipzig, and the Oregon petitions, Congress refused to ratify the Kyoto. However, in signing the Leipzig, I was baptized into a battle I would be forced to fight within my own domain of the newsbiz, as I quickly discovered my colleagues were officially in the tank for global warming. Tragically, their bias has burgeoned over the years.

In their quest to push their anthropogenic agenda, the Society of Environmental Journalists (SEJ), an affirmed and respected organizational arm of the popular media, has assembled a global warming guide which includes a list of "Skeptics and Contrarians."

According to the SEJ:

> Some of the most vocal skeptics have done relatively little recent peer-reviewed scientific research on the topic, and some have had their voices amplified via financial support from industries opposed to any government regulation or taxation of greenhouse gas emissions ... overall, their number represents a distinctly minority position in the ongoing and normal colloquy among scientists about the evidence of climate change and its likely impacts.[182]

---

182 Society of Environmental Journalists website, "Climate Change, a Guide to Information and Disinformation," http://www.sej.org/resource/index18.htm.

The SEJ website then names names, providing a *de facto* blacklisting service on whom not to call, especially attempting to link their list of "skeptics and contrarians" to the oil industry and capitalism. In turn, the SEJ provides a lengthy contact list of "trusted" activists, including James Hansen, who would "be glad to talk to a journalist," the website offers.

Interestingly, engaging in a bit of investigative journalism of my own, I discovered that the SEJ has received considerable funding from radical groups well-known for their enthusiastic support to supposedly save the earth from man's CO_2.

According to tax returns buried on their website, the SEJ accepted largess from the Rockefeller Family Fund, an organization whose "environment program emphasizes public education on the risk of global warming . . . and enforcement of the nation's environmental laws. . . ."[183] Also noteworthy is the money received from the Turner Foundation (as in Ted Turner, founder of CNN). The Turner Foundation has assisted a host of environmental groups over the years and is best known for its pledge of $1 billion to the United Nations.

In a wild appearance on PBS's Charlie Rose Show on April 1, 2008, Ted Turner predicted that because of global warming, "We'll be eight degrees hotter in 30 or 40 years, and basically none of the crops will grow . . . most of the people will have died and the rest of us will be cannibals." Turner's outrageous comments on Rose's program were not meant to be an April Fool's joke. He continued to describe our climate-altered future: "Civilization will have broken down. The few people left will be living in a failed state—like Somalia or Sudan—and living conditions will be intolerable."

Turner's solution? Right from the Ehrlich playbook—*stabilize* the population. He told Rose, "We're too many people—that's why we have global warming." He continued, "Too many people are using too much stuff" and "on a voluntary basis, everybody in the world's got to pledge to themselves that one or two children is it."

The SEJ's credibility was further trashed at their annual convention in October 2009, following a speech by their keynote guest, Al Gore. Assuming he was on friendly turf, following his diatribe there was a very

183 Rockefeller Family Fund website, http://www.rffund.org/environment/.

rare Q and A session. By chance, among those able to ask a question was Irish filmmaker Phelim McAleer, director of *Not Evil, Just Wrong*, a movie critical of the global warming movement.

Stepping to the audience microphone, McAleer queried Gore about the 2007 decision by the British High Court regarding *An Inconvenient Truth*. McAleer asked, "The judge in the British High Court, after a lengthy hearing, found that there were nine significant errors [in the movie]. . . . This has been shown to children. Have you—do you accept those findings, and have you done anything to correct those errors?"

Gore adeptly sidestepped the question by going straight to the heart-strings of the useful idiots assembled in the room, responding, "The ruling was in favor of the movie by the way, and the ruling was in favor of showing the movie in the schools and that [sic] that's that's really [sic] the the bottom line on that. There's been such a long discussion of each one of those, uh, specific things, um. One of them for example was that polar bears—if I remember correctly, it's been a long time ago—that polar bears really aren't endangered. Well, polar bears didn't get that word so, uh . . ."

McAleer countered by saying, "Well, the number of polar bears have increased, and actually are increasing."

"You don't think they're endangered, do you?"

"The number of polar bears have increased."

"Do you think they're endangered?"

"The number of polar bears have increased."

At this point the stacked house began to jeer McAleer, who attempted to continue above the hostile audience. "I mean, if the number of polar bears increase, surely they're not endangered. And the judge, a judge did have a lengthy hearing . . . no, but no, I mean—"

McAleer was cut off by former SEJ President Tim Wheeler of the *Baltimore Sun*. "We're not doing a debate here," Wheeler said.

"No," McAleer pressed, "it's a question. He hasn't answered the question—"

McAleer's microphone was immediately shut down.

Now that's some good, old-fashioned journalism—just like the kind formerly employed in the Soviet Union.

Uh, and about those polar bears: an enlightening paragraph from a

May 2007 story in the *Christian Science Monitor,* dealing with the polar bear population boom stands out:

> "There aren't just a few more bears. There are a . . . lot more bears," biologist Mitchell Taylor told the *Nunatsiaq News* of Iqaluit in the Arctic territory of Nunavut. Earlier, in a long telephone conversation, Dr. Taylor explained his conviction that threats to polar bears from global warming are exaggerated and that their numbers are increasing. He has studied the animals for the Nunavut government for two decades.[184]

The Society of Environmental Journalists and the propagandists they assist in the mainstream media are nothing but the feeding trough for sloppy science, brewed in caldrons of bias, and stirred by statist politicians and elitist social engineers. And sucking it up like chubby piglets are thousands of useful idiots.

SHUT UP AND ACT WARM

Hans Christian Andersen's "The Emperor's New Clothes" is a charming story of an idiot king who provides swindlers with a large sum of money in exchange for a new royal suit of clothes. The swindlers pretend to work feverishly on a loom for days fashioning the clothing, when in fact they are doing nothing but planning on how to spend their treasure. As ministers of the royal court visit to check on their work, the fraudulent tailors point to an empty fabric loom and confidently proclaim, "Is not the material beautiful?"

Embarrassed by their own lack of vision, the ministers foolishly reply, "Is not the cloth magnificent? See here, the splendid pattern, the glorious colors."

After his own review of the clothiers' fictitious work, the puzzled emperor thinks to himself, "This is terrible. Am I so stupid? Am I not fit to be emperor? This is disastrous." But, because apparently everyone in his court can see the new garments, he says aloud, "Oh, the cloth is perfectly wonderful. It has a splendid pattern and such charming colors."

---

184 Fred Langon, "Canadian Controversy: How do polar bears fare?" *Christian Science Monitor,* May 3, 2007.

Eventually, the new "outfit" is worn by the emperor. Prancing about blindly in the buff, he prepares for a parade.

His subjects have heard much about the new wardrobe and line the streets to see the procession.

"Oh, how splendid are the emperor's new clothes. What a magnificent train! How well the clothes fit," the fools cried.

But among the crowd a little child suddenly gasps, "But he hasn't got anything on."

If that kid were around today, he would laugh himself silly at the dissonance of the global warming chorus.

Enter the swindlers who sold the U.N. a climate conference in Montreal, Canada, in early December 2005, which ended up coinciding with a record cold snap. Six thousand activists and global warming propeller-heads attended.

The city was in the grip of a winter come early. Below-average temperatures in the 20s blew through Montreal, yet no one dared bat an eye to the obvious irony of their shouts that the earth was about to combust.

As the conference began, thousands of useful idiots took to the streets donned in parkas, woolen hats, and even "dressed as polar bears to urge leaders at a U.N. climate conference to do more to curb global warming," according to the *Sydney Morning Herald*.[185]

The *Herald* continued, "'Time is running out' banners proclaimed in a carnival-like rally in freezing temperatures through central Montreal, where many protesters accused the United States of blocking progress on climate change and threatening the world's future."

While the frozen masses were preoccupied with trying to pretend they were warm, Elizabeth May of the Sierra Club shouted through a P.A. system, "We will move the world ahead!"

Cue Hans Christian Anderson's little child: "Wouldn't a little global warming would be a *good thing* about now?"

It gets even more blind.

On February 13, 2007, much of the United States was being blasted with a dose of biting Arctic air. Minneapolis, Minnesota, woke up to −4° temps. Chicago had snow and a temperature of 19°. So ferocious was the

---

185 "Thousands March Over Global Warming," *The Sydney Morning Herald*, December 4, 2005.

storm that Maryville College in St. Louis cancelled its screening of Al Gore's *An Inconvenient Truth*.

As the storm quickly raced east, Washington, D.C. was bracing for its biggest snowfall of the season. Alarmed by the weather forecast for the following day, a notice was fired off to BlackBerrys across Capitol Hill: "The Subcommittee on Energy and Air Quality hearing scheduled for Wednesday, February 14, 2007, at 10:00 a.m. in room 2123 of the Rayburn House Office Building has been postponed due to inclement weather."

The title of the scheduled hearing was "Climate Change: Are Greenhouse Gas Emissions from Human Activities Contributing to a Warming of the Planet?"

The cancellation announcement was beheld by the world on the Committee's website (see Figure 7.1).

Figure 7.1: Cancellation of Hearing by House Subcommittee on Energy and Air Quality

The Senate had also planned a big climate change shindig, and, despite the House of Representatives' move to bag their global warming meeting because of *non*global warming conditions, the Senate moved forward with theirs. Foreign bigwigs had been called to Washington for this minisummit, and canceling it would have created great inconvenience. As the dignitaries cruised from their D.C. hotels through the snow-covered city in four-wheel-drive Suburbans, they witnessed the fluffy white evidence of the biggest snowfall of the season. In addition, the temperature was a stunning 11° *below* normal.

Undaunted by the wintry reality outside, stubborn global warming believer, Senator John McCain, addressed the gathering, and without flinching said, "The debate is over, my friends. The question is, what do we do?"

Oh, I can hear that little kid yelling out at the top of his lungs, "Put some clothes on, Senator! You look ridiculous in the buff!"

If this laughable scenario was not the height of foolishness and hypocrisy, I am hard-pressed to find another.

I honestly believe McCain is one of those who, unlike Gore, Hansen, and the current crew in the White House, really believes the earth is warming, or, perhaps, the climate is changing and it's all man's fault. McCain has bought the lie and somehow believes he can help solve this supposed crisis.

But for the love of God, Senator, just walk down the hallway and spend some time with some of your fellow Republican senators who are responding to the science, not the propaganda.

A GREAT AMERICAN

When it's time to move out of California, I just might consider Oklahoma. In the Sooner state there are still more pickup trucks than Priuses, more heroes than hippies, and more statesmen than statists representing the people in Washington. Leading the charge in the fight against the climate liars is Oklahoma Senator James Inhofe. Sitting opposite Barbara Boxer as the minority leader on the Senate's Environment and Public Works Committee, Inhofe has compiled a superb list of climate professionals who dispute anthropogenic global warming claims. The impressive international roster now includes some 800 scientists and professionals who work in the field—not kooky activists working for socialist, nongovernmental organizations.

Included on Senator Inhofe's roster is one of the dozen people who walked on the moon: Dr. Harrison Schmidt (Apollo 17, 1972). In 2009, Schmidt, a Ph.D. geologist, shocked many by resigning from the highbrow Planetary Society that was founded by Carl Sagan. The society's stated mission is "to inspire the people of Earth to explore other worlds, understand our own, and seek life elsewhere." Predictably, this organization has dived headfirst into the global warming tank.

Unable to subject himself to such folly, Schmidt, who in the 1970s also served as a U.S. Senator, presented a resignation statement:

As a geologist, I love Earth observations. But, it is ridiculous to tie this objective to a "consensus" that humans are causing global warming when human experience, geologic data and history, and current cooling can argue otherwise.

"Consensus," as many have said, merely represents the absence of definitive science. You know as well as I, the "global warming scare" is being used as a political tool to increase government control over American lives, incomes and decision making. It has no place in the Society's activities.[186]

Another prominent scientist to sign on to Inhofe's report is Joanne Simpson, the first woman in the world to receive a Ph.D. in meteorology. The preeminent Dr. Simpson, formerly of NASA, has authored more than 190 scientific studies. She, too, has had enough: "Since I am no longer affiliated with any organization nor receiving any funding, I can speak quite frankly.... As a scientist I remain skeptical.... The main basis of the claim that man's release of greenhouse gases is the cause of the warming is based almost entirely upon climate models. We all know the frailty of models concerning the air-surface system."[187]

Inhofe also received the endorsement of many scientists who worked in lead capacities for the United Nation's Intergovernmental Panel on Climate Change, including nuclear physicist and chemical engineer, Dr. Philip Lloyd of South Africa. Dr. Lloyd has authored over 150 publications, and says regarding global warming: "The quantity of CO_2 we produce is insignificant in terms of the natural circulation between air, water and soil. ... I am doing a detailed assessment of the U.N. IPCC reports and the Summaries for Policy Makers, identifying the way in which the Summaries have distorted the science."

The senator's all-star list includes a winner of the Noble Prize for Physics, Dr. Ivar Giaever, who stated: "I am a skeptic ... Global warming has become a new religion."

And the compilation of brain power reflects the thoughts of a former

---

186 "Ex-Astronaut: Global Warming is Bunk," Associated Press, February 16, 2009.
187 U.S. Senate Minority Report, Senator James Inhofe, Senate Committee on Environment and Public Works, updated March 17, 2009.

member of Greenpeace, Jarl R. Ahlbeck, a chemical engineer at Åbo Akademi University in Finland and author of more than 200 scientific publications: "So far, real measurements give no ground for concern about a catastrophic future warming."

Additionally, Senator Inhofe also has signed many representatives from the community of hurricane researchers, a body that defies the notion of consensus. Government scientist Stanley B. Goldberg of the National Oceanic and Atmospheric Administration's (NOAA) Hurricane Research Division, said: "It is a blatant lie put forth in the media that makes it seem there is only a fringe of scientists who don't buy into anthropogenic global warming."

In fact, the supposed hurricane connection to global warming and associated climate change is, as most of the global warming arguments are, laughable.

HURRICANE CONSENSUS

In May 2008, the country of Myanmar was in the direct path of a cyclone (hurricanes in the southern hemisphere are known as cyclones). A solid "Category 3," on a scale of 1 to 5 when it made landfall, the cyclone unfurled winds of over 120 miles per hour and surged ocean water into the low-lying, heavily populated coast for many miles. Even though the path of the storm had been accurately forecast for many days, an effective notification system had not been instituted by the communist government of Myanmar. As a result, the death toll is estimated to have been 100,000.

Al Gore did not blame the repressive Myanmar government, or even the weather, for this horrible loss. Rather, he blamed you and me.

On the May 6 edition of National Public Radio's *Fresh Air*, Gore explained to host Terry Gross, "And as we're talking today, Terry, the death count in Myanmar from the cyclone that hit there yesterday has been rising from 15,000 to way on up there to much higher numbers now being speculated." In a tone and cadence that sounded as if the interviewer was in kindergarten, Gore continued, "And last year a catastrophic storm last fall hit Bangladesh. The year before, the strongest cyclone in more than 50 years hit China—and we're seeing consequences that scientists have long predicted might be associated with continued global warming."

And then Gore had the gall to play the consensus card.

"It's also important to note that the emerging consensus among the climate scientists is that even though any individual storm can't be linked singularly to global warming, we've always had hurricanes," Gore said. "Nevertheless, the trend toward more Category 5 storms—the larger ones—and the trend toward stronger and more destructive storms appears to be linked to global warming and specifically to the impact of global warming on higher ocean temperatures in the top couple of hundred feet of the ocean, which drives convection energy and moisture into these storms and makes them more powerful."

And Terry Gross soaked up Gore's flood of false science without the slightest challenge.

If you want to observe a branch of meteorology where there is virtual consensus regarding global warming, look no farther than to those who actually study and forecast hurricanes. This community of scientists knows Gore is prancing about *au naturel*.

Re-enter Dr. William Gray, unquestionably the world's foremost hurricane forecaster. He founded the Tropical Meteorology Project at Colorado State University in the 1960s, where he developed the fine art of forecasting hurricanes in the Atlantic Basin, including the Gulf of Mexico. Numerous times Dr. Gray has told my radio audience, "I am of the opinion that global warming is one of the greatest hoaxes ever perpetrated on the American people."

Gray's opinion is not based on a gut feeling—it's based on the science. And he is not alone.

"All my colleagues that have been around a long time—I think if you go to ask the last four or five directors of the National Hurricane Center—we all don't think this is human-induced global warming," says Dr. Gray.[188]

Indeed, another pioneer in hurricane research, and a 13-year director

---

188 Interview with Dr. William Gray on KSFO, San Francisco, April 7, 2007.

of the National Hurricane Center, Dr. Neil Frank, told the *Washington Post*, "It's a hoax."[189]

When asked if he thought increasing carbon dioxide in the atmosphere could actually be a good thing he replied, "Exactly! Maybe we're living in a carbon dioxide-starved world. We don't know."[190]

COUNTING HURRICANES

When the Gulf of Mexico coastline took its direct hit from Hurricane Katrina in 2005, a columnist for the *Boston Globe* blamed the storm on global warming. "The hurricane that struck Louisiana yesterday was nick-named Katrina by the National Weather Service. Its real name is global warming," he wrote.[191]

Actually, if the buffoonish columnist wanted to accurately name Katrina, he would have taken off his tin-foil cap and called the storm "normal weather."

Hurricanes are the fascinating *grandes dames* of earth's natural weather machine. To pin their frequency and intensity on global warming is fool-ish, but to the uninformed, it's a sexy sell.

In 2004 and 2005, the global whiners received costly gifts from the global warming gods. The 2004 hurricane season saw a near record six hurricanes striking U.S. soil. Of those, a record three were classified as major hurricanes (Category 3 or higher). In 2005, six hurricanes also made landfall in the United States, with a new mark of four reaching major status—the most devastating, of course, being Katrina.

While it may seem compelling to link these natural disasters to global warming, recall from Chapter Three that 2005 was the 16th hottest year on record, and 2004 didn't crack the top 20. Historically, these two devastat-ing hurricane seasons need to be placed in proper perspective (Table 6.1):

---

189 Joel Achenbach, "The Tempest," *Washington Post*, Sunday, May 28, 2006.
190 Ibid.
191 Ross Gelbspan, "Katrina's real name," *Boston Globe*, August 30, 2005.

TOTAL HURRICANES STRIKING UNITED STATES		MAJOR HURRICANES STRIKING UNITED STATES	
Number	Year	Number	Year
7	1886	4	2005
6	1916, 1985, 2004, 2005	3	1893, 1909, 1933, 1954,
5	1893, 1909, 1933		2004
4	1869, 1880, 1887, 1888,	2	1879, 1886, 1915, 1916,
	1906, 1964		1926, 1944, 1950, 1955,
3	31 years have 3 strikes		1985

Table 6.1: Hurricanes Striking United States

The left half of this chart indicates that 2004 and 2005 both experienced six hurricanes strikes in the United States. If global warming was to blame for those busy seasons, how do you explain the equal number of hurricanes that came ashore in 1916, one of the colder years on record? And could the record *seven* strikes in 1886 be caused by man's burning of fossil fuels? Obviously not; it was nearly as cold as 1916.

The right side of the chart indicates that 2005 holds the record of four major hurricanes hitting our shores. However, one must ask if an unnatural increase in CO_2 is to blame. What was driving the three major storms that hit the United States in 1893 and 1909? Again these were very chilly, post–Little Ice Age years that have no link to carbon dioxide. Admittedly, 1954 was a hotty, but our proof-by-association exercise is running out of gas.

It should also be noted that despite the frequency and intensity of the hurricanes during the 2004–2005 period, the following two hurricane seasons were tame. No hurricanes touched U.S. soil in 2006, and 2007 would also have been hurricane-free were it not for the last gasps of Humberto, which limped into Texas with winds of 90 mph. All told, 2006 produced a mere five hurricanes in the Atlantic Basin (the lowest since 1997) and 2007 only six—kind of a bummer if you're in the hurricane research biz, and even more so if you're using hurricanes to raise money to fight global warming. The year 2008 was a bit more active, tallying eight hurricanes, with remnants of Hurricane Ike coming ashore in Galveston as a relatively weak Cat-2. But 2009 was a bust for hurricane hunters, with the term producing only three Atlantic hurricanes, making it the most lackluster season since 1997.

"But, Brian," the worldly global whiner might interrupt, "you're so Yankee-focused. How can you judge the hurricane seasons based on how many storms hit the United States? The Atlantic Ocean and surrounding waters are a huge breeding ground for hurricanes."

Well, let's check the stats (Table 6.2):

HURRICANES IN ATLANTIC BASIN		MAJOR HURRICANES IN ATLANTIC BASIN	
Number	Year	Number	Year
15	2005	8	1950
12	1969	7	1961, 2005
11	1887, 1916, 1950, 1995	6	1916, 1926, 1955, 1964,
10	1870, 1878, 1886, 1893,		1996, 2004
	1933, 1998	5	1893, 1933, 1951, 1958,
9	1880, 1955, 1980, 1996,		1969, 1995, 1999
	2001, 2004		

Table 6.2: Hurricanes in the Atlantic Basin

To the left, we see that 2005 was an anomaly with a record 15 hurricanes forming in the Atlantic. The Summer of Love experienced the second greatest number (and no, you can't blame it on what was smoked at Woodstock, nor can it be blamed on temperature—1969 was a cool year). Following those records, we see many years with a significant number of hurricanes, half of which occurred in the 1800s.

The right side of the chart illustrates that 1950 (another cool year) holds the record for the most powerful storms, with eight major hurricanes. Beyond that, we see a smattering of years with intense storms, most having played out prior to 1960.

Again, to try to pin frequency and intensity on global warming is folly. Like all kinds of weather, hurricanes simply happen. On average, close to seven hurricanes every four years (1.8 per year) strike the United States, while about two major hurricanes cross the U.S. coast every *three* years.

Consider some other noteworthy hurricanes, none of which occurred in particularly hot years:

Deadliest Hurricane: More than 8,000 people perished September 8, 1900, when a Category 4 hurricane barreled into Galveston, Texas. The storm surges exceeded 15 feet, and winds howled at 130 mph, destroying more than half of the city's homes.

Most Intense Hurricane: An unnamed storm slammed into the Florida Keys during Labor Day, 1935. Researchers estimated sustained winds reached 150–200 mph with higher gusts. The storm killed an estimated 408 people.

Greatest Storm Surge: In 1969, Hurricane Camille produced a 25-foot storm surge in Mississippi. Camille, a Category 5 storm, was the strongest storm of any kind to ever strike mainland America. When the eye hit Mississippi, winds gusted up to 200 mph. The hurricane caused the deaths of 143 people along the coast from Alabama into Louisiana and led to another 113 deaths as the weakening storm moved inland.

Earliest and Latest Hurricanes: The hurricane season is defined as June 1 through November 30. The earliest observed hurricane in the Atlantic was on March 7, 1908, while the latest observed hurricane was on December 31, 1954. The earliest hurricane to strike the United States was Alma, which struck northwest Florida on June 9, 1966. The latest hurricane to strike the United States was on November 30, 1925, near Tampa, Florida.

Hurricanes, one of the favorite proofs that advocates of anthropogenic global warming use to validate their claims, have become earth's biggest bogeyman.

Reporters can call names, senators can make unfounded pronouncements, the Terminator can pump his biceps, and the U.N. can hold conferences on impending doom, but the only consensus regarding the connection of hurricanes and global warming is that there is no connection between the two.

However, like slick slip-and-fall lawyers, the Marxist elites pushing their social engineering agenda are not about to let a few facts thwart their plans. There's too much wealth at stake that needs to be spread around.

8

PLAYBOOK

We should change our attitude toward the United Nations. There has to be some power in the world superior to our own.

—Andy Rooney, *60 Minutes*

WHEREAS THE MISSION of the United States, as penned in the Declaration of Independence, promises to secure the rights to "Life, Liberty and the pursuit of Happiness," the mission of the United Nations seeks to "promote social progress." The rights proclaimed in the former are unalienable, or God-given. The latter pursues a theoretical scheme concocted by imperfect humans who actually believe that, by way of Marx's Law of Transformation, they are best suited to determine how their fellow man shall live.

Because the right to own personal property is a foundational principal of liberty and the pursuit of happiness—and is antithetical to theories ascribed to Karl Marx—the overwhelming body of the United Nations has always despised the United States and has sought an effective means of reshaping her.

After observing the wild success of the first Earth Day in 1970, U.N. bureaucrats surmised their most effective tool to whittle away the United States' primary pillar of liberty would be the environment. Accomplishing this goal would require a masterful playbook, a willing fifth column in the media and, ultimately, the opportune team in the White House.

In 1976, the U.N. boldly unleashed their first move to emasculate Uncle Sam by introducing their playbook.

PLAY ONE: LAND GRAB

Play number one was officially articulated at the United Nations Conference on Human Settlements, or Habitat I, held in Vancouver, British Columbia, in 1976. Agenda item 10 of the Conference Report presented the U.N.'s new official policy regarding land. The Preamble states:

> Land . . . cannot be treated as an ordinary asset, controlled by individuals and subject to the pressures and inefficiencies of the market. Private land ownership is also a principal instrument of accumulation and concentration of wealth and therefore contributes to social injustice; if unchecked, it may become a major obstacle in the planning and implementation of development schemes. The provision of decent dwellings and healthy conditions for the people can only be achieved if land is used in the interests of society as a whole. Public control of land use is therefore indispensable. . . .

Clearly, Karl Marx could not have articulated this key portion of Habitat I more effectively. To the United Nations, private land ownership is a mechanism of social injustice; it's akin to criminal, and therefore must be dealt with. And notice another vital element within the Preamble: private property ownership "if unchecked . . . may become a major obstacle in the planning and implementation of development schemes." The nefarious minds at the U.N. believe that until property owners are vanquished, they will have difficulty moving forward with their program of global transformation.

To effectively begin to punish property owners and institute social justice, Habitat I introduced a devious device: taxation. Quoting from Habitat's agenda item 10:

> Excessive profits resulting from the increase in land value due to development and change in use are one of the principal causes of the concentration

of wealth in private hands. Taxation should not be seen only as a source of revenue for the community but also as a powerful tool to encourage development of desirable locations, to exercise a controlling effect on the land market and to redistribute to the public at large the benefits of the unearned increase in land values.[192]

By progressively taxing land owners, the U.N. sought to force four outcomes: a source of revenue for government welfare programs; a forsaking of private property (because of the high tax burden); control over the real estate market; and the blatant redistribution of wealth from money that was perceived as having been "unearned."

The recommendations brought forward in Habitat I were agreed to by President Gerald Ford's U.N. ambassador, William Scranton. However, like the majority of proposals that flow from the bowels of the United Nations, Habitat I was considered "soft law"—a nonbinding, strategic tactic designed to keep testy constituents back home appeased. However, by endorsing the recommendations crafted in Vancouver, Scranton was discreetly informing the world body, "We're with you."

PLAY TWO: SUSTAINABLE DEVELOPMENT

By the 1980s, Earth Day had evolved into a traditional institution, and, as hoped, the constant drumbeat noting the critical harm humans were supposedly inflicting upon the earth's ecosystem had turned more and more Americans into guilt-ridden, eco-wimps. Ready to capitalize on this weakness, the United Nations and their comrades, discreetly holed-up on American soil, were patiently waiting to execute their next move to further the recommendations of Habitat I. All they had to do was wait out the presidency of Ronald Reagan.

Reagan, however, instinctively saw through the socialist/communist shenanigans of the U.N., as noted by his selection of Jeane Kirkpatrick as ambassador to the United Nations. Kirkpatrick was a no-nonsense

---

192 United Nations Conference on Human Settlements (Habitat I), Agenda Item 10, Recommendation D.3 (a).

whistle-blower who was willing to stand up to the pomp and poppycock that, to this day, permeates the 38-story U.N. megaplex overlooking the East River in midtown Manhattan.

Once, in a bold move to illustrate the hypocrisy of the faux U.N. peace-niks, Kirkpatrick stood before the world body and read names from a long list of political prisoners held in the gulags of the Soviet Union. She was confronted with cold stares; after all, the Soviet Union was an upstanding member of the United Nations and one of five countries making up its influential Security Council.

Sadly—but true to her promise—Jeane Kirkpatrick served only one term, stepping down in 1985. Knowing Congress would not confirm another firebrand like Kirkpatrick as U.N. ambassador, Reagan selected a less enthusiastic appointee, Vernon Walters, to take her place for the last years of his presidency.

Relieved and giddy that Kirkpatrick was out of the way and Reagan would soon be termed out of office, the U.N., in 1987, trotted out their next play, cloaked in a newly defined, Orwellian term: "sustainable development."

Having made clear their beliefs regarding private property and social justice in Habitat I, the concept of sustainable development would allow the emboldened U.N. leadership to heap additional layers of autocratic theory onto the playing field. They would also mesmerize the public by renaming the earth "the biosphere," refer to the transfer of wealth as "economic growth," turn energy rationing into "lifestyle adjustment," and specifically link carbon dioxide to a new enemy, "global warming."

It was all announced in the Bruntland Report, commonly referred to as Our Common Future, published by the United Nations' World Commission on Environment and Development. In Chapter One, "sustainable development" is officially defined:

> The concept of sustainable development does imply limits—not absolute limits but limitations imposed by the present state of technology and social organization on environmental resources and by the ability of the biosphere to absorb the effects of human activities . . . sustainable

development requires meeting the basic needs of all and extending to all the opportunity to fulfill their aspirations for a better life.[193]

Meeting essential needs requires not only a new era of economic growth for nations in which the majority are poor, but an assurance that those poor get their fair share. . . .[194]

Sustainable global development requires that those who are more affluent adopt life-styles within the planet's ecological means—in their use of energy, for example.[195]

Later, in the document, CO_2 is officially introduced as a global demon that threatens world peace:

With the exception of CO_2, air pollutants can be removed from fossil fuel combustion processes at costs usually below the costs of damage caused by pollution. However, the risks of global warming make heavy future reliance upon fossil fuels problematic. . . .[196]

All nations may suffer from the releases by industrialized countries of carbon dioxide and of gases that react with the ozone layer, and from any future war fought with the nuclear arsenals controlled by those nations. All nations will also have a role to play in securing peace, in changing trends, and in righting an international economic system that increases rather than decreases inequality, that increases rather than decreases numbers of poor and hungry.[197]

Sustainable development was the new commandment; masterfully tying the environment to the economics of social justice. The faceless goons who crafted *Our Common Future* were hopeful that this new prod would be their most effective instrument yet to bring about global societal and political change. The key would be getting the United States on

---

193 *Our Common Future: Report of the World Commission on Environment and Development,* Chairman's Forward, subchapter 3.27, United Nations, March 20, 1987.
194 Ibid., sub sub chapter 3.28.
195 Ibid., sub sub chapter 3.29.
196 Ibid., Chapter 7, sub chapter 2.18.
197 Ibid., Chapter 12.3.

board, because—just as in any major battle, or even a minor bar fight—once the biggest guy goes down, a winner is born.

PLAY THREE: AGENDA 21

In the five years following the Brundtland Report, the world's elites and policymakers rallied around the new flagpole of sustainable development. Working feverishly behind the scenes, the next play from their tyrannical treatise was crafted and eventually rolled out with unprecedented global fanfare at the 1992 United Nations' Earth Summit, in Rio de Janeiro, Brazil. Arranged "to help governments rethink economic development," the summit was the largest conference of its kind ever held. Virtually every country on the planet was represented, with 108 heads of state personally attending—including the lackluster, politically struggling, President George H.W. Bush. The number of slobbering journalists covering the event numbered nearly 10,000, with seemingly none critical of the summit's centerpiece: *Agenda 21*.

Agenda 21 was presented—and is still regarded—as the social engineer's bible. The Agenda's preamble states:

> Humanity stands at a defining moment in history. We are confronted with a perpetuation of disparities between and within nations, a worsening of poverty, hunger, ill health and illiteracy, and the continuing deterioration of the ecosystems on which we depend for our well-being. However, integration of environment and development concerns and greater attention to them will lead to the fulfillment of basic needs, improved living standards for all, better protected and managed ecosystems and a safer, more prosperous future. No nation can achieve this on its own; but together we can—in a global partnership for sustainable development.

Whereas the Brundtland Report introduced the world to new terminology and provided a diagnosis of the planet's supposed ills, Agenda 21 details the envisioned cures. The sweeping scheme articulates plans to end poverty, describes adequate housing as a "right,"[198] and urges each country provide its citizens with universal health care.[199]

Agenda 21 further clarifies the goals of sustainable development by pushing countries toward creating enforceable "environmental laws and regulations that are based upon sound social, ecological, economic and scientific principles"[200] and "the need to control atmospheric emissions of greenhouse and other gases and substances"[201] as well as "promote appropriate energy efficiency and emission standards."[202] Schools are to be utilized to "encourage education and awareness-raising programmes at the local, national, subregional and regional levels concerning energy efficiency."[203]

Proving he was no Reagan, and certainly not a conservative, the first Bush actually signed Agenda 21. Although legislation was never passed by Congress to mandate the Agenda, it didn't matter. Five months after Rio, one-termer Bush lost the White House to Bill Clinton and Al Gore, two politicians who would carry the sustainable development baton more adeptly than Bush ever could.

According to Vice President Gore, sustainable development would achieve a "sacrifice, struggle, and a wrenching transformation of society."[204]

Emboldened by the newly elected progressive White House administration, on March 29, 1993, San Francisco's tilt-left Congresswoman Nancy Pelosi submitted legislation urging the federal government to "assume a strong leadership role in implementing the decisions made at the Earth Summit developing a national strategy to implement Agenda 21 and other Earth Summit agreements."[205]

Her legislation failed to pass, perhaps because her colleagues feared

---

198 *Agenda 21*, United Nations Environmental Program, Chapter 7.6.

199 Ibid., Chapter 6:13.

200 Ibid., Chapter 8.14.

201 Ibid., Chapter 9.9.

202 Ibid., Chapter 9.12 j.

203 Ibid., Chapter 9.12 k.

204 Al Gore, *Earth in the Balance* (Plume, 1993), 274 (paperback version, there was no way I was going to spend more money to purchase the hardcover edition, which was originally published in 1992).

205 H.J. Resolution 166, Rep. Nancy Pelosi, March 29, 1993.

supporting Pelosi might harm them in the coming fall congressional elections. However, once again, it didn't matter. The wishes of the future House Madam were granted three months later.

PLAY FOUR: U.S. IMPLEMENTATION

On June 29, 1993, Pelosi's intention was fulfilled, and a huge portion of America's liberty stealthily torn asunder, when President Clinton signed Executive Order 12852. The EO established the president's Council on Sustainable Development and simultaneously appointed Vice President Gore to conduct a National Performance Review. There was no vote of the legislature and no pesky day-after headlines to deal with, just, as Clinton advisor Paul Begala later bragged to the *New York Times*, a "Stroke of the pen. Law of the land. Kind of cool."[206]

The President's Council seated 25 members including most of his cabinet secretaries, representatives from radical environmental groups such as The World Resources Institute, Natural Resources Defense Council, and the Women's Environment and Development Organization, plus a few token representatives from industry. Working in tandem with Gore's National Performance Review, an overhaul of the Departments of Interior and Agriculture were conducted to implement what Gore called the "Ecosystem Management Policy." This new protocol was coordinated with existing legislation, such as the Endangered Species Act and the Clean Water Act, authorizing the federal government supreme power to regulate land use in rural America. Additionally, the Ecosystem Management Policy provided state and local governments with financial incentives to aggressively regulate land use, and it implemented another subscheme from Agenda 21: stakeholders.

Prior to 1993 the term "stakeholder" usually referred to gambling; the stakeholder was the person in charge of holding the money put up for a wager. But via one sentence buried within Agenda 21, the noun had taken on new meaning:

---

206 James Bennet, "True to Form, Clinton Shifts Energies Back to U.S. Focus" *New York Times*, July 5, 1998.

Encourage the principle of delegating policy-making to the lowest level of public authority consistent with effective action and a locally driven approach.[207]

The reference to "lowest level of public authority" does not exclusively refer to elected city councils or county commissions. Instead, this clause instituted civilian councils that work with the government in public-private partnerships. These stakeholder councils have become quite mainstream and are generally headed by a representative committed to an eco-oriented, Non-Governmental Organization (NGO) like the Nature Conservancy, Sierra Club, World Wildlife Fund, or Greenpeace, allowing them a sanctioned seat at the table without ever having to be elected into any office. The duty of the stakeholders is to create public pressure to enforce and create local, state, and federal environmental laws; raise private money; and garner further public funding—all with the ultimate goal of declaring large swaths of land off-limits to development. Stakeholder councils have been known to use shakedown tactics, abuse eminent domain, conjure claims of endangered species, and procure phony scientific studies to achieve their desires.

Stakeholders, and their like-minded co-conspirators working within all levels of government, commonly refer to a manual created by Gore's Ecosystem Management Policy: the *Growing Smart Legislative Guidebook: Model Statutes for Planning and the Management of Change*. *Growing Smart* provides detailed instructions on how to create legislation, implement tax policies, and use existing laws and regulations to squeeze private landowners. Tactics from the guidebook have provided the conspiratorial template to assist in the creation of needless open-space reserves, forest preserves, marine reserves, and supposed ecologically sensitive areas—all with the goal of removing mass acreage from future private development.

Taking it a step further in 1995 the President's Council on Sustainable Development presented a report entitled, "Sustainable America, A New Consensus." The report arrogantly boasts a number of "We Believe" statements, which clearly are the antithesis of the foundations of American liberty:

207 Ross Gelbspan, "Katrina's real name," *Boston Globe*, August 30, 2005.

We Believe,

Economic growth, environmental protection, and social equity are linked. We need to develop integrated policies to achieve these national goals.[208]

The United States should have policies and programs that contribute to stabilizing global human population. . . .[209]

Even in the face of scientific uncertainty, society should take reasonable actions to avert risks where the potential harm to human health or the environment is thought to be serious or irreparable.[210]

Did any of the manifest propeller-heads who wrote this statement ask *you*, if this is what *you* believe?

When this council speaks of our national goals being embodied in the "integration of economic growth, environmental protection and social equity," they don't speak for the majority of Americans; they represent the psyche of the sock-it-to-America tyrannists, who plot from a 38-story building overlooking the East River.

When these elitists speak of "stabilizing global human population," they mean activating the diabolic machinations of Erhlich and Holdren. Abortion, rampant disease in the third world, mass casualties of war, nonbreeding homosexuality, and the societal peer pressure for the traditional family to bear only 1.5 children are all tolerable solutions to these egalitarian goons.

And proving they are fronting the biggest scam in history, these brazen, global whining cons are willing to create sweeping social policy "even in the face of scientific uncertainty."

They have also cleverly created mechanisms to silence their opposition. One of the 154 action items in the New Consensus report states: "The federal government should play a more active role in building consensus on difficult issues."[211] In other words, procure bogus studies, sell

---

208 President's Council on Sustainable Development, *Sustainable America—A New Consensus*, We Believe Statement, number 10, The White House, 1995.

209 Ibid., Number 11.

210 Ibid., Number 12.

211 *Sustainable America: A New Consensus for the Prosperity, Opportunity and a Healthy Environment for the Future*, Natural Resources Stewardship, Policy Recommendation 1, Action 4, President's Council on Sustainable Development, 1995, Chapter 5.

them in the schools and through the media, marginalize the opposition, and always remember: the end justifies the means.

PLAY FIVE: 2009 STIMULUS BILL

Habitat I, The Brundtland Report, Agenda 21, and the Clinton/Gore initiatives for implementation laid a mythical utopian foundation for the next significant play in the United Nations' attempt to impose social justice. With their utilization of phony science and *ad hominem* arguments to pressure nescient drones into believing that the carbon dioxide created from the use of fossil fuels will destroy life on this planet, the U.N. may have made their boldest move yet, the United Nations' Framework Convention on Climate Change (UNFCCC).

Secretly crafted throughout 2008 during the run-up to the presidential elections, and publicly presented two months after Barack Obama took office in 2009, the UNFCCC dictates include enacting new "policies and measures that are purely domestic in scope (taxes, levies, subsidies, policy reform, public investment, cap and trade regimes, technology mandates)" as well as "international agreed taxes and levies, [and] internationally agreed cap and trade regimes."[212]

Given that Europe has already foolishly instituted taxes, levies, subsidies, policy reforms, and CO_2 cap and trade; realizing that Russia does whatever it pleases; knowing that Africa has no significant industry; understanding that Latin America always receives a free pass; and noting that prior soft law and treaties exclude China and India; the UNFCCC was drafted with the United States solely in its crosshairs. This becomes crystal clear when one reads further into the document and discovers the call for "green public investment" and "public funding for environmentally sound energy infrastructure and smart grids"

---

212 From document FCCC/KP/AWG/2009/INF.3, Section III, released March 16, 2009, as prepared for the United Nations Framework on Climate Change, Ad Hoc Working Group On Further Commitments For Annex I Parties Under The Kyoto Protocol, 7th Session, Bonn, Germany, March 29–April 8, 2009.

(to be explained in the following chapter) to be implemented in developed nations.[213]

It must be revealed that all of the stated UNFCCC directives were fulfilled in the gargantuan, $787 billion, 2009 Congressional "stimulus" bill, deceptively entitled the American Recovery and Reinvestment Act, which represented the largest one-time spending act in the history of the world. I often wondered aloud on my radio program who took the time to write this 1,000-plus page document (let alone who had the time to read it). Clearly the un-American NGOs were conspiring with their U.N. comrades.

The commensurate UNFCCC-payouts found in the Congressional stimulus bill includes:

- $20 billion in funding for demonstration programs and basic research, much of which relates to energy efficiency and environment.
- $14.9 billion for energy efficiency and conservation programs at the state and local levels, and for loan guarantees for renewable energy projects.
- $16 billion for the weatherization of private homes occupied by lower-income Americans. An additional $300 million would provide rebates for the purchase of energy-efficient home appliances.
- $25 billion targeted for renovations to federally owned buildings and facilities. The bulk of this funding will be deployed in accordance with ridiculous green building design and construction protocols created by the federal government for its own buildings.
- $20.8 billion for environmental cleanups, wastewater treatment, watershed reclamation, and national park and forest restoration.
- $13.3 billion for green vehicle and mass transit programs, including rail and light rail.

---

213 Ibid.

- $11 billion for modernizing the nation's power grid (i.e., the Big Brother "smart grid").

PLAY SIX: A COMRADE IN THE WHITE HOUSE

Say what you want about the second President Bush—"W"—but you can't claim he was a fan of the United Nations. His ambassador of choice to that body of loons was John Bolton, a man who clearly recognized the Marxist marrow of the U.N. Bolton was so threatening that congressional Democrats prevented him from gaining official confirmation to the post, forcing Bush to assign him on a recess appointment.

Though his stay at the U.N. lasted less than two years, Bolton picked up where Kirkpatrick left off, stating, "There is no such thing as the United Nations. There is only the international community, which can only be led by the only remaining superpower, which is the United States."[214] It was Bolton who also jested that the building which houses the U.N. "in New York has 38 stories. If you lost ten stories today, it wouldn't make a bit of difference."[215]

Proving that elections have consequences, currently we have not just the antithesis of John Bolton representing our interests in the United Nations, but we have an entire team surrounding our president who may as well be running the entire show in that 38-story den of iniquity. And, for many of us, this is no surprise: we knew where Barrack Obama's allegiances lay well before he was elected to the White House.

Our first unmistakable sign of Obama's aspirations to carry water for the U.N. occurred while he was on the campaign trail. Totally unreported by the mainstream media, in July 2008 Senator Obama co-sponsored Senate Bill 2433—the Global Poverty Act. Reading from the *Congressional Record*, the act:

---

214 Roland Wilson, "Bush deploys hawk as UN envoy," *Times* (London), March 8, 2005.
215 Anne Applebaum, "Defending Bolton," *Washington Post*, March 9, 2005.

Directs the President, through the Secretary of State, to develop and implement a comprehensive strategy to further the U.S. foreign policy objective of promoting the reduction of global poverty, the elimination of extreme global poverty, and the achievement of the United Nations Millennium Development Goal . . . [by] 2015.

Before proceeding, we must briefly unpack the Millenium Development Goals (MGD). The goals were made public in 2000, agreed to by the vast majority of countries and heralded by former-President Bill Clinton. The problem for fans of the MGD was that "W" never saw fit to adopt the promises proclaimed in the document—promises that include eradicating poverty, broader access to free abortions, integrating "the principles of sustainable development into country policies and programmes,"[216] and to "develop further an open, rule-based, predictable, nondiscriminatory trading and financial system."[217]

Knowing that he was likely to be anointed president of the United States, Obama's sponsorship of the Global Poverty Act was a sign to one-worlders that he was their guy . . . and Bill's wife Hillary, who by this time had been thrown under the bus in the election process for Obama, would be their gal as Secretary of State.

More specifically, Obama's Poverty Act would require the United States to achieve specific and measurable goals consistent with those established by the United Nations Millennium Resolution, including:

. . . mobilizing and leveraging the participation of businesses and public-private partnerships; coordinating the goal of poverty reduction with other internationally recognized Millennium Development Goals; [and] integrating principles of sustainable development and entrepreneurship into policies and programs.[218]

---

216 *The Millennium Development Goals Report 2008,* published by the United Nations Department of Economic and Social Affairs, August 2008, p. 36.
217 Ibid., p. 44.
218 Senate Bill 2433, "Global Poverty Act," Senator Barack Obama, lead sponsor, introduced December 7, 2007.

A press release from then-Senator Obama's office declared:

> In 2000, the U.S. joined more than 180 countries at the United Nations
> Millennium Summit and vowed to reduce global poverty by 2015. We are
> halfway towards this deadline, and it is time the United States makes it a
> priority of our foreign policy to meet this goal and help those who are
> struggling day to day.[219]

Jeffrey Sachs, who runs the "Millennium Project," confirms the U.N.
plan to force the United States to pay 0.7% of GNP (in current dollars
that would be about $75 billion, on top of the $25 billion we presently
allocate overseas). The only way to raise that funding, Sachs confirms, "Is
through a global tax, preferably on carbon-emitting fossil fuels."[220]

Not accounting for inflation, Senate Bill 2433's aim was to throw $750
billion dollars into this money pit over the next ten years, and, as always,
the United States would have no discretion regarding how the taxpayer
dollars would be spent.

Interestingly, Achim Steiner, who heads the U.N. Environmental
Program (UNEP), shared with world leaders at the 2009 G-20 summit
in London, a report from his office suggesting that donations of just 1%
of a country's gross domestic product could bankroll a "Global Green
New Deal" to the coincidental tune of about $750 billion annually. He
even floated the option of taxing the oil consumption of rich nations to
fund the transfer of wealth scheme. "If, for argument's sake," he said, "you
were to put a five year levy in OECD countries of $5 a barrel, you would
generate $100 billion per annum." Minimizing the outlay, he added, "It
translates to roughly 3 cents per liter."[221]

The American arm of the Global Green Deal was revealed in the massive 1,201-page global warming bill originally introduced by Congressmen

---

219 Larry Petrash, "Reject Obama's Senate Bill S. 2433," Times Record News, August 3,
2008.
220 Vincent Gioia, "United Nation's Power Will Grow Under Obama," Post Chronicle,
December 31, 2008.
221 Alister Doyle, "$750 billion 'green' investment could revive economy: U.N." Reuters,
March 19, 2009.

Henry Waxman from southern California and Ed Markey from Massachu-setts in May, 2009 and passed by the House of Representatives in June. Given the warm and fuzzy title, "The American Clean Energy and Security Act of 2009," the monster missive reads like a damned U. N. resolution.

Initially, I may have been one of the few in America who actually read the bill. I was so outraged by its contents that I emailed popular syndi-cated radio host and author Mark Levin, who immediately responded by inviting me on his show. Ten minutes later I was speaking to his seven mil-lion listeners, sharing the socialist giveaways buried in the bill. Section 2201 describes the "Energy Refund Program." This portion of the bill enables those having an income of up to 150% of the poverty level (about $30,000) to be eligible for a yet to be determined "energy refund," which will be "pro-vided in monthly installments via direct deposit into the eligible household's designated bank account." In addition to this blatant transfer of wealth, the legislation brashly contains a similar international redistribution component modeled after the aforementioned United Nations Framework Convention on Climate Change. Section 441 of the legislation actually reads:

> Under Article 4 of the United Nations Framework Convention on Climate Change, developed country parties, including the United States, "take all practical steps to promote, facilitate, and finance, as appropri-ate, the transfer of, or access to, environmentally sound technologies and know-how to other parties, particularly developing country parties . . . the United States [is] committed to enhanced action on the provision of finan-cial resources and investment to support action on mitigation and adapta-tion and technology cooperation. . . . "

The American Clean Energy and Security Act of 2009, as I shared with Mark's audience, is a promise to utilize climate change as a tool to redistribute the hard-earned money of achievers with those on the lower economic rung domestically; and internationally the Act will transfer bil-lions of dollars to the third world, supposedly to create "environmentally sound technologies and know-how." However, instead the money will be skimmed by U.N.-approved middlemen, spent without accountability by a variety of non-governmental organizations, and pocketed by dictators and warlords—just as our money funneled abroad always is.

PLAY SEVEN: DREAM TEAM

Like a dream come true for those beholden to the goals of the United Nations, Barrack Hussein Obama was sworn in to office January 20, 2009 (actually, he was sworn in privately on the 21st, since the initial oath of office was flubbed during the public ceremony—and the second time around there was no Bible on which to place his hand), and in his first address as president promised to "restore science to its rightful place." The restoration process began in earnest with the selections of his teammates. Not only do all acknowledge the U.N. as a legitimate world body, but all are totally in the tank regarding the wealth redistribution, global warming nexus. These radical elites are best represented by their own words:

> **Susan Rice,** Ambassador to the United Nations, in her introductory speech to the U.N. stated, "To tackle global warming, all major emitting nations must be part of the solution. . . . And we should help the most vulnerable countries adapt to climate change."[222]

> **Hillary Clinton,** during her Senate confirmation hearings said, ". . . climate change is an unambiguous security threat. At the extreme, it threatens our very existence."[223]

> **Lisa Jackson,** Administrator, Environmental Protection Agency, adds a racial component to her beliefs, "Climate change is a clear and present danger to communities of color across the country. . . . We must act now to not only end, but also reverse, the ravaging effects inflicted upon our homes, schools and neighborhoods."[224]

---

222 Susan Rice, remarks made during presentation of her credentials to UN President Ban Ki-moon, January 26, 2009.
223 Hillary Clinton, Senate Confirmation Hearings, January 13, 2009.
224 Michael H. Cottman, "Lisa Jackson: Blacks Should Get Serious About Climate Change" blackamericanweb.com, February 16, 2009, http://www.blackamericaweb.com/?q=articles/news/moving_america_news/6877.

Nancy Sutley, Chairwoman of the President's Council on Environmental Quality, stated during her Senate confirmation hearing that climate change "is an issue that will affect the entire federal government, almost no agency is untouched by climate change."[225]

Carol Browner, Energy Czar, believes "we face an environmental, a public health, and an economic challenge in global warming and climate change unlike anything we have faced thus far."[226]

Stephen Chu, Secretary of the Department of Energy, has said, "Climate change . . . will cause enormous resource wars, over water, arable land, and massive population displacements. We're talking about . . . hundreds of millions to billions of people being flooded out, permanently."[227]

Hilda Solis, Secretary of Labor, "I am fighting for environmental justice."[228]

John Holdren, Director of the White House Office of Science and Technology Policy, "We are already experiencing 'dangerous anthropogenic interference' with the climate system. . . . To fix the problem, society has only three options: mitigation, adaptation, and suffering. We're already doing some of each, and will do more of all three."[229]

---

225 Nancy Sutley, remarks before the Environmental and Public Works Committee, during EPA confirmation hearings, January 14, 2009.

226 Carol Browner, "Crisis of Climate Change" lecture, Clinton School Speaker's Series, University of Arkansas, November 11, 2006.

227 Comments by Stephen Chu at National Clean Energy Summit convened by the University of Nevada Las Vegas, Senator Harry Reid (D-NV), and the Center for American Progress Action Fund, Summer 2008. Speech available for viewing at YouTube, "Dr. Stephen Chu at National Clean Energy Summit."

228 Remarks delivered by California State Senator Hilda L. Solis at the John F. Kennedy 2000 Profile in Courage Award Ceremony, May 22, 2000.

229 John Holdren, speaking at the Kennedy School of Government, November 6, 2007, http://www.climatesciencewatch.org/index.php/csw/detailsgetting_to_know_holdren _part_2/.

Ken Salazar, Secretary of Interior, at his confirmation hearing, "There is no doubt that climate change and global warming is having an impact on a whole host of natural features of this world, including endangered species that we have."[230]

Tim Geithner, treasury secretary, told members of Congress, "If people don't change how they use energy then they will face higher costs for energy."[231] Cass Sunstein, regulatory czar, in a 2007 paper entitled, "Climate Change Justice," says, "... if wealthy people in wealthy nations want to help poor people in poor nations, emissions reductions are far from the best means by which they might to do so," and "when people in one nation wrongfully harm people in another nation, the wrongdoers have a moral obligation to provide a remedy to the victims. It might seem to follow that the largest emitters, and above all the United States, have a special obligation to remedy the harms they have caused."[232]

With the playbook written and the dream team assembled, the transformation of America is well underway. As of this writing, the American Clean Energy and Security Act of 2009 is still being debated in the Senate. However, because the players in this seditious conspiracy are longsuffering, if they can't get what they want in a single, sweeping chunk of legislation, they'll obtain it later in smaller bites, by tucking components of their plan into various appropriations bills, or more likely, via a combination of Executive Orders.

Though a right-minded majority in Congress would be a help in obstructing this gut-wrenching transformation, we desperately need an authentic conservative in the White House; one who will institute sound science and old-fashioned common sense to dismantle these regressive

230 Jim Tankersley, "Ken Salazar promises reform at Interior Department," *Los Angeles Times,* January 16, 2009.
231 Treasury Secretary Tim Geithner, speaking before the House of Representatives Budget Committee, March 3, 2009.
232 Eric A. Posner, Cass R. Sunstein, "Climate Change Justice." The Law School, University of Chicago, August, 2007.

environmental strategies and put America back on a track toward abundant energy resources.

In the meantime, Obama and crew press forward with their plans to suppress America's energy resources, gain further control of our lives, and make a few of their faithful exceedingly rich.

Better cling to your guns and religion, because, as you know, this mob has plans.

9

NO ALTERNATIVES

I can't understand why there aren't rings of young people blocking bull-dozers, and preventing them from constructing coal-fired power plants.

—Al Gore

THE WHITE HOUSE Office of Science and Technology Policy (OSTP) was established by Congress in 1976. Its mission is to serve "as a source of scientific and technological analysis and judgment for the President with respect to major policies, plans and programs of the Federal Government."[233] While certain past OSTP Directors could certainly be described as political liberals, none could be so easily tagged an environmental radical as Barack Obama's OSTP Director, John Holdren can.

Holdren's unorthodox belief system is well-known and accepted among environmentalists, and his selection by Obama speaks volumes regarding the content of the president's character and his plans to radically transform America. Holdren has written: "Indeed, it has been concluded that compulsory population-control laws, even including laws requiring compulsory abortion, could be sustained under the existing Constitution if the population crisis became sufficiently severe to endanger the society."[234] Even

---

233 U.S.C. tit. 42, ch. 79, subch. II, sec. 6614.
234 John Holdren, Anne Ehrlich, and Paul Ehrlich, *Ecoscience; population, resources, environment* (San Francisco: W. H. Freeman and Company, 1977), p. 837.

creepier, Holdren has penned, "Adding a sterilant to drinking water or staple foods is a suggestion that seems to horrify people more than most proposals for involuntary fertility control.... To be acceptable, such a substance would have to meet some rather stiff requirements: it must be uniformly effective, despite widely varying doses received by individuals, and despite varying degrees of fertility and sensitivity among individuals...."[235] In a 1995 paper, Holdren explained his model for sustainable development, noting "humans are included as just one species and are not treated specially."[236]

In my Foreword I included another Holdren citation that speaks of his obvious disregard for the lifestyle advances achieved by the people of the United States: "A massive campaign" Holdren says, "... to restore a high-quality environment in North America and to de-develop the United States," must occur. To achieve this end "resources and energy must be diverted from frivolous and wasteful uses in overdeveloped countries to filling the genuine needs of underdeveloped countries."[237]

By "de-develop" Holdren has further explained he means "lower per-capita energy consumption, fewer gadgets, and the abolition of planned obsolescence."[238]

Obama's chief scientist is cut from a distinct brand of U.N. cloth that perceives humanity from a Marx-meets-Darwin-and-Malthus viewpoint. For those of us who hold an opposing worldview and do not see humans as being on par with a cockroach, it's difficult, if not impossible, to comprehend where such thinking comes from, but know this: the Holdrens of the world are dead serious about de-developing America, particularly through the creation of reduced-carbon energy policies that will limit our available energy supply and retard our way of life. And don't be fooled by their speeches about alternative energy resources—it's a ruse. These extremists desire less energy, not more; after all, "humans are included as just one species and are not treated specially."

---

235 Ibid., pp. 787–788.
236 John Holdren, Gretchen C. Daily, and Paul Ehrlich, "The Meaning of Sustainability: Biogeophysical Aspects."
237 John Holdren, Anne Ehrlich, and Paul Ehrlich, *Human Ecology: Problems and Solutions* (San Francisco: W.H. Freeman and Company, 1973), p. 279.
238 John Holdren and Paul Ehrlich, "Introduction," in Holdren and Ehrlich, eds., *Global Ecology,* 1971, p. 3.

THE ANSWER IS NOT BLOWING IN THE WIND

It was interesting to note that in the Democrats' $787 billion "stimulus bill" there were little more than token giveaways for wind and solar enthusiasts. With all the money being freely printed, one would have thought at least a few billion could have been doled out to erect electricity-producing wind farms in the plains and pastures of solar panels in the desert. Instead, the struggling wind and solar industry received not cash, but a multitude of complex tax credits and subsidies. It was the donkey party's way of appeasing their faithful and allowing political allies who have invested in wind and solar to make a few easy bucks.

I'm not denying the validity of wind power—those turbines really do spin out energy. However, currently less than 1% of our nation's electricity is generated via wind, and it's doubtful that percentage will ever rise significantly—for two reasons. First, wind power is wholly unreliable. Advocates can place turbines wherever they want, but when the breeze stagnates, electricity production halts. Conversely, when the wind blows too hard, the massive turbine blades are forced to shut down, lest they bend inward and snap the mast like a pencil, causing the entire structure to dangerously shred apart (search YouTube and you'll witness evidence of several wind turbines blowing up during storms[239]). Thus, wind is, at best, a supplemental form of energy that requires a full-time backup power source.

Secondly, wind power is not compatible with the demands of hardcore environmentalists because it requires the development of broad swaths of land. For example, a thought-provoking illustration I often suggest compares the real-estate footprint of a nuclear plant to that of a wind farm.

The Comanche Peak nuclear power plant outside Dallas, Texas, is a significant facility that produces about 2,300 megawatts of power—more than enough to serve the electricity needs of 1.3 million average-sized homes. The plant fits neatly into 8,000 acres and includes a large reservoir used for cooling the plant, which can also double as a source of recreation. Compare that landmass to the one required for the highly publicized Pampa Wind Project, promoted by Dallas hedge fund manager, T. Boone

---

239 My personal "favorite" turbine failure is found at http://www.youtube.com/watch?v=7nSB1SdVHqQ.

Pickens. The highly touted "Pickens Plan" envisioned supplying power to an equal number of homes, but required 400,000 acres of Texas real estate.[240] Besides erecting thousands of massive masts upon which the turbines are fixed, Pickens' plan necessitated the construction of transmission towers and lines and associated service roads.

In an obvious reaction to the drumbeat of environmental protestors, in July 2009, Pickins quietly dumped his Pampa project, hoping to perhaps construct a few significantly smaller wind farms in Kansas, Oklahoma, and Wisconsin. Too bad for T. Boone. If he had simply talked to anyone in the San Francisco Bay Area, he could have saved the estimated $60 million he spent trying to get his Pampa plan off the ground.

In the 1970s, just east of the Bay, the world's largest concentration of wind turbines was constructed. Some 4,500 windmills are ensconced atop 50,000 acres of grassy hills, presently generating a modest 576 megawatts of power. Officially known as the Altamont Pass Wind Resource Area, one would suppose the wind farm is an icon of greenness. But instead, Altamont Pass is the poster girl of eco-infighting.

Ever since the multitude of three-bladed rotors was installed, a significant increase in the numbers of dead birds in the area have been reported. Animal rights activists immediately went ballistic, demanding action. Since then, lawsuits have been filed, and millions of dollars spent procuring studies to track the bird body count in an effort to determine how to address the problem. In 2008, the most extensive study was released: a two-year, taxpayer-funded examination conducted by the Altamont Pass Avian Monitoring Team. It surveyed 2,500 of the turbines and kept meticulous records of the bird body count. During the study period, 1,596 rotary-blade bird deaths were confirmed, including the deaths of 633 raptors.[241] Extrapolating their data to account for all 4,500 windmills on the farm, as well as estimates regarding how many dead birds the researchers didn't count before scavengers made off with an easy meal, the monitoring team claims

---

240 *Pickens orders $2 billion in wind turbines,* Associated Press, May 15, 2008.
241 "Bird Fatality Study at Altamont Pass Wind Resource Area, October 2005–September 2007," Altamont Pass Avian Monitoring Team, Draft Report, January 25, 2008.

that in two years 8,247 birds died of what I refer to on the radio as "turbine to the head syndrome."

"We are deeply distressed about the continuing bird deaths," said Elizabeth Murdock, executive director of the Golden Gate Audubon Society, and chief plaintiff in a lawsuit that has shaped the costly war between the animal rights zealots and their alternative energy counterparts.[242]

The bird lovers are winning this battle, and it's unlikely new turbines will be constructed at Altamont Pass anytime soon, let alone in many other locations around the country.

NO SOLUTION IN SOLAR

Solar power is a much more realistic alternative than wind, and I've personally utilized it to reduce my own home's electric bill since the Eighties. However, once again, eco-activists are the solar alternative's biggest enemy.

In 2005, San Diego Gas & Electric (SDG&E) signed a joint initiative with Phoenix-based Stirling Energy Systems. Their mutual goal was to harness 900 megawatts of solar power in the harsh, barren Southern California desert. The plan seemed like a natural one: sunny skies 90% of the time in the middle of nowhere. Vital to this proposal, however, was the construction of a $1 billion high-power transmission line to transport the energy to the end-users. It took no time for critics to react, loudly voicing their disapproval of the environmentally-damaging infrastructure that would cut right through the Anza-Borrego Desert State Park. Three years after the initial planning began, an administrative law judge from the California Public Utilities Commission, Jean Vieth, stopped the project, arguing that the 150-foot high towers would harm species in the park. The judge called the potential impact of the solar project "frightening."[243]

---

242 Charles Burriss, "The Deadly Toll of Wind Power," *San Francisco Chronicle,* January 2, 2008.

243 Syanne Olson, "The end of the road for San Diego Gas & Electric's 900MW renewable energy project?" *Daily News,* PVtech.org, , November 8, 2008.

Judge Vieth also stated that the California Public Utility Commission's global warming policy objectives "do not justify the severe environmental damage that any of the transmission proposals would cause."[244]

Not surprisingly, no reporters of this eco-drama ever revealed the back-story of what was really occurring: Anza-Borrego Desert State Park, located in the Mojave Desert, is protected by an agreement with the United Nations that declares the region a World Biosphere Reserve. There are 525 such reserves around the world, 47 in the United States, 30 of which are managed by the U.S. National Park Service. According to the U.S. Department of State, each biosphere reserve—like Anza Borrego—is to serve three functions: "conservation of important biological resources; development of environmentally sound economic growth; [and] support of environmental research, monitoring, education."[245]

It's the phrase "development of environmentally sound economic growth" that should raise suspicion. Translated it means: *no one will benefit economically if it means disturbing the environment.* This was confirmed in a *Los Angeles Times* headline that proclaimed that powerful California Senator Diane Feinstein "wants desert swaths off-limits to solar, wind projects."[246]

Indeed, it was Senator Feinstein who helped cobble together much of land that is protected by the biosphere reserve. In 1994 Feinstein was the primary sponsor of what she considers to be one of her proudest achievements, the California Desert Protection Act. In a plan straight out of Agenda 21, working with stakeholder groups, more than $40 million was raised to buy former railroad land in the Mojave Desert, turning it over to the government in one of the largest land purchases in California history. Once the transactions were complete, the famed Death Valley was immediately transformed into a 3.4 million acre national park (the largest in the lower 48 states), and the 1.6 million acre Mojave National Preserve was established. These parcels were combined with Anza-Borrego, Joshua

---

244 Ibid.

245 Senate Bill Report HJM 4029, as reported by Senate Committee On Natural Resources and Parks, February 20, 1998.

246 Richard Simon, "Feinstein wants desert swaths off limits to solar, wind projects," *Los Angeles Times,* March 25, 2009.

Tree National Monument, and the Santa Rosa Mountains Wildlife Area to be crowned a U.N. biosphere reserve.

So, when speaking to the *Times* about thwarting efforts to build a solar facility in the middle of a protected biosphere reserve, Lady Di wasn't fooling when she said, "I feel very strongly that the federal government must honor that commitment."[247]

So, is it any wonder that of the 130 solar projects that have applied for permits with the Department of the Interior, to build and operate power generating facilities in the Mojave, none have yet to obtain one?

While solar is an efficient alternative, unless the arrays are constructed on private property don't expect the sun to be adding significantly to our overall energy supply.

HYDRO A NO-NO

With all the hype about wind and solar, which together generate less than 2% of America's energy, 10% of our electricity comes from water. Hydroelectric power is 100% pollution free and creates no carbon dioxide. It uses the kinetic energy of rushing water to make electricity. Most often dams are constructed to stop the flow of a river, with the water behind the dam forming a beautiful reservoir. The collected water is gravity fed through a large pipe called a penstock. The flow from the penstock pushes against blades in a turbine at the base of the dam, causing them to spin and generate clean, efficient electricity. In California, we receive 15% of our power from hydro, while Washington State generates about 85% of its electricity from such facilities, and even exports surplus power to nearby states.

But once again, the environmentalists have targeted yet another green alternative with the Sierra Club even launching a "Harmful Hydro" campaign. The antihydro environmentalists believe the reservoirs drown trees and displace salamanders. And when these activists fail to pull heartstrings with the "submerged species" ruse, they exaggerate and falsify claims of fish dying downstream from the dams.

---

247 Ibid.

As usual, their propaganda efforts have been largely effective, and in recent years hundreds of small hydroelectric dams have been demolished. In 2007, at least $17 million was spent to destroy the 47-foot-tall Marmot Dam in Oregon. It was the largest dam yet to be decommissioned. The Marmot is said to have generated nearly $6.5 million dollars worth of electricity annually for the Portland General Electric Company (PGEC), but, after caving to the pressure of well-organized stakeholder councils, the reliable source of green energy was removed. Sounding like a hostage reciting a ransom note, PGEC President Peggy Fowler said, "This partnership took a great step toward restoring a breathtaking river for fish, wildlife, and people."[248]

Given that the Marmot had been in place for the better part of a century, fish and wildlife had long since adapted to the effects of the dam, the people in and around Portland had enjoyed an inexpensive source of electricity, and Portland General's shareholders had received a nice return on their investment. Removing the dam was a terrible step in the wrong direction, because now, emboldened by their success in Oregon, the no-hydro hacks are placing substantial pressure all across America to tear down dozens of larger dams used to generate substantial loads of power, even though the replacement energy would be generated from CO_2-producing coal or natural gas.

In fact, in California, a collection of stakeholder groups are gaining serious traction to destroy the O'Shaughnessy Dam, a massive engineering marvel approved by an act of Congress in 1913 and located in a remote area of the Sierra Nevada mountains, 160 miles east of San Francisco. The 312-foot-tall dam is fed by snow runoff and supplies the water needs of roughly two million residents of the Bay Area, while also generating 500 megawatts of power to several hundred thousand California homes. My prediction is, despite the Golden State arguably having the most expensive water and electricity in the country, the whiners will eventually win this battle with no alternative power or water source created in the dam's stead. There never are alternatives or compromises with the environmentalist activists, just the constant demand for the rest of us to use less damn energy.

---

248 Bernie Woodal, "Utility blasts Oregon dam to make way for fish," Reuters, July 24, 2007.

NO NUKES

Nuclear power is another carbon-free source of electricity, but one adamantly opposed by the same clan. Their primary scare tactics are Three Mile Island and Chernobyl. While the 1986 Chernobyl, Ukraine, nuclear plant explosion was a horrendous, deadly disaster created by incompetent communist bureaucrats in the former Soviet Union, the 1979 Three Mile Island breakdown has been revised to look like a calamity. While Three Mile Island did suffer a severe core meltdown (the most dangerous kind of nuclear power accident), only minor off-site amounts of radioactivity were released, with no deaths or injuries to plant workers or residents of the nearby community. Essentially, the plant shut down exactly the way it was designed to do in the event of malfunction.

Nonetheless, the accident at Three Mile Island was used to generate permanent public fear and mistrust, which has been milked by the anti-nuke crowd ever since. As a result, since 1979, there have been no new, nonmilitary, nuclear facilities constructed in the United States, though many existing plants have withstood lawsuits and protests and have been completed or expanded since.

Currently there are 104 reactors, supplying 20% of America's electricity. Sans the protestors, if we had continued to expand our nuclear infrastructure at the pace proposed in the 1960s and 1970s, we could be more like France, which generates nearly 80% of its electricity with nuclear technology.

While the first argument against nuclear power is a reactor meltdown, the second polemic is nuclear waste, and both have been used by President Obama. During the primaries, when he was still being refined and navigating his way through the national press, he recklessly exposed his anti-nuke notions, telling a New Hampshire newspaper's editorial board, "I don't think there is anything we inevitably dislike about nuclear power. We just dislike the fact that it might blow up, and irradiate us, and kill us! That's the problem."[249] The fawning media clucks chuckled.

---

249 Barack Obama, during a meeting with the Editorial Board of the Keene Sentinel newspaper in New Hampshire, on Nov. 25, 2007, and reported in "Nuclear Power a Thorny Issue For Candidates," National Public Radio *Morning Edition,* July 21, 2008, by David Kestenbaum. Available on youtube.com "Sen. Barack Obama on nuclear power from SentinelSource.com."

Currently, all nuclear waste created by all of America's nuclear plants since the 1960s, are stored at 126 different sites scattered about the nation. The total amount of spent uranium (the primary waste material) amounts to 57,000 tons. Determining it would be best to safeguard these nuclear byproducts in one location, the U.S. Department of Energy, in 1978, began studying a potential repository at Yucca Mountain, Nevada, about 100 miles northwest of Las Vegas. I used to live in Nevada. It's a massive state with topography resembling the rugged, barren terrain of Afghanistan. Apart from a handful of population centers that are home to just over two million residents, Nevada is a dry, dusty, tumbleweed-ridden, no-man's land—the perfect place to secure nuclear waste.

Granted, 57,000 tons of anything sounds like a lot, but that amount of spent uranium does not require an expansive storage area. Extremely dense, a chunk of uranium the size of a gallon of milk weighs 150 pounds, illustrating that only 13 gallon-jugs of uranium would roughly equal a ton. Simple math would break down the uranium tonnage into roughly 760,000 milk jugs, all which could fit neatly within the confines of the average high school basketball gym.

Candidate Obama made it clear he was against Yucca Mountain, and once president, his first proposed budget cut off money for the nuclear waste repository, meaning that the $10 billion in taxpayer money, spent since 1983 to ready Yucca for storing spent nuclear waste, was wasted.[250]

But at the end of the day, for the elites, nuclear and monetary waste is not the issue—inexpensive, plentiful power is. John Holdren's writing partner, Paul Ehrlich, summed it up, saying, "Giving society cheap, abundant energy . . . would be the equivalent of giving an idiot child a machine gun."[251]

---

250 "Obama dumps Yucca Mountain" *World Nuclear News,* February 27, 2009, http://www.world -nuclear-news.org/newsarticle.aspx?id=24743.

251 "An Ecologist's Perspective on Nuclear Power," *Federation of American Scientists Public Issue Report,* May/June 1978.

DEATH OF OL' KING COAL

Obama's propensity to pop off about energy issues was a glaring problem when in the company of liberal newspaper editorial boards during
his campaign. In a January 17, 2008, interview with the editors of the *San
Francisco Chronicle*, candidate Obama divulged, "... if somebody wants to
build a coal-powered plant, they can; it's just that it will bankrupt them
because they're going to be charged a huge sum for all that greenhouse
gas that's being emitted."[252] Proving their Obama bias, the *Chronicle* conveniently omitted his quote in the next day's edition.

Electricity derived from coal powers 50% of all homes in the United
States. Like nuclear power, coal is a relatively inexpensive energy resource,
and its supply is abundant. The United States has the world's largest
known coal reserves, an estimated 489 billion tons, enough to last hundreds of years at today's level of use.[253] Coal is natural in 27 states and
the coal industry directly employs over 170,000 blue-collar workers.[254]

Yes, there *are* potentially harmful pollutants (*not* CO_2) associated
with coal, but since the Sixties, through superior technological advances
applied in the United States, dangerous impurities such as sulfur and
nitrogen oxides, and particulates (soot) have been reduced by 90% even
though the use of coal has tripled.[255] Nonetheless, regardless of this veri-
fiable coal clean-up, Obama's Energy Secretary, Steven Chu, has repeatedly said, "Coal is my worst nightmare."[256]

Like Holdren, despite the laid-back demeanor, Chu's a major radical, and if he gets his way, every coal plant in the country is in jeopardy
of being closed. Fully aware that soot, sulfur and nitrogen oxides are no

---

252 "Audio: Obama Tells SF Chronicle He Will Bankrupt Coal Industry," *Newsbusters.org*,
reported by P.J. Gladnick, November 2, 2008. Audio available at youtube.com "SHOCK
Audio Unearthed OBAMA TELLS SAN FRANCISCO HE WILL BANKRUPT THE
COAL INDUSTRY."
253 United States Energy Information Administration, "Coal Reserves," February 2009.
254 Ibid., "Coal Mining Productivity by State and Mine Type," September 2008.
255 John Entine, "Coal and Climate Change—Can King Coal Clean Up?" American
Enterprise Institute, March 9, 2009, posted at www.aei.org.
256 Keith Johnson, "Steven Chu: 'Coal is My Worst Nightmare,'" *Wall Street Journal*,
December 11, 2008.

longer a problem associated with coal in the United States, his carping focuses on two bogus targets—CO_2 and a certifiable red herring known as "fly ash."

In a coal-fired power plant, coal is burned to create steam, which, in turn, spins turbines which generate electricity. Once the coal is consumed, two different types of ash remain. One is a rather dense byproduct that falls to the bottom of the burning chamber and is appropriately called "bottom ash." The other is very light and, in days gone by, escaped into the air, creating soot and pollution. That's called fly ash. In the 1930s, technology was brilliantly developed to capture the fly ash and recycle it. Today, fly ash is often recycled as a primary ingredient in concrete, stucco, and other mortar products, giving these construction materials amazing strength.

Besides the recycling factor, one would think the greenies would be completely supportive of recycling fly ash in concrete, because recycling it reduces the amount of water and cement that would otherwise be used to create the product, and less mixing means less CO_2 expelled during the process. By selling fly ash to concrete manufacturers, utility companies also lower their costs, which passes savings on to the consumer—a sure win-win for everybody.

Fly ash technology and use is such a no-brainer that the California Department of Transportation requires mineral admixtures, including fly ash, to comprise at least 25% of the cement material used in any state-funded paving projects.

Currently nearly one-third of the fly ash created in the United States is used in concrete manufacturing. The remainder is disposed of in properly engineered landfills and abandoned mines. Fly ash is *not* a problem, but do you think that would give Chu pause? Unfortunately, not a chance.

"COAL IS REALLY BAD"

Prior to taking the helm as secretary of energy, Steven Chu was a keynote speaker at the 2007 World Affairs Council of Northern California's Climate Change and Global Politics Conference. Having a great interest in how elites think, and given that the conference was held just down the street from my radio studio, I observed Chu's speech, notepad in hand. Though he didn't repeat his patent "worst nightmare" line, he did whine that "coal is really bad":

> Coal plants are polluting in very many different ways, and just so you can get a little bit easier with nuclear power, a coal plant that captures 99.5% of the fly ash, which is radioactive, emits approximately 100 times more radioactivity than a nuclear power plant. And the fly-ash is radioactive and that's assuming that that fly ash is safely disposed of, okay? It's actually not safely disposed of, it's put in roads, it's put in other things, it leaks into the water table as well. So, we're digging this stuff out of the ground and spreading it over the surface of the earth—something to think about. Coal is really bad.[257]

Comparing fly ash to the radioactivity of a nuclear plant is a deceptively clever tactic designed to simultaneously deride both coal and nuclear power. At issue is coal's uranium and thorium content, both naturally radioactive elements occurring in such trace amounts that they pose absolutely no health risk. In making his skewed argument, Chu is banking on his audience not being able to recall specifics from their ninth grade geology class, namely, that everything emanating from the earth possesses *some* amount of radioactivity. Consequently, Chu's claim immediately tickles the ears of ignorant eco-liberals.

However, according to the United States Geological Survey (USGS), which maintains what many regard as the largest database involving the chemical composition of coal in the world:

> Radioactive elements in coal and fly ash should not be sources of alarm. The vast majority of coal and the majority of fly ash are not significantly enriched in radioactive elements, or in associated radioactivity, compared to common soils or rocks[258] [and] measurements of fly ash radioactivity

---

257 Stephen Chu, "A New Energy Program," speech presented to the World Affairs Council, San Francisco, September 13, 2007.
258 "Radioactive Elements in Coal and Fly Ash" USGS Fact Sheet FS 163-97, October 1997.

leached into the ground and water "indicate the dissolved concentrations of these radioactive elements are below levels of human health concern."[259]

Proving he's a clever activist, Chu's contention that fly ash is 100 times more radioactive than a nuclear plant is intentionally misleading, because given that the amount of radiation emitted by a nuclear plant is virtually undetectable to the surrounding population, 100 times *nothing* equals *nothing.* His bogus comparison was popularized in a headline from a 2007 *Scientific American* article, entitled "Coal Ash Is More Radioactive than Nuclear Waste."[260] However, digging into the details of the piece, one reads a quote from Dana Christensen, associate lab director for energy and engineering at the esteemed Oak Ridge National Laboratory, proving that health risks from radiation in coal by-products are low. Christensen says, "Other risks like being hit by lightning are three or four times greater than radiation-induced health effects from coal plants."

The *Scientific American* article further debunks Chu's case, stating:

> So why does coal waste appear so radioactive? It's a matter of comparison: The chances of experiencing adverse health effects from radiation are slim for both nuclear and coal-fired power plants—they're just somewhat higher for the coal ones. "You're talking about one chance in a billion for nuclear power plants," Christensen says. "And it's one in 10 million to one in a hundred million for coal plants."

NOT IN CHU'S BACKYARD

In Chu's San Francisco speech, he also addressed carbon sequestration. This is a proven technology that takes the CO_2 created during the burning of coal, liquefies it, and pumps it into the ground. Though a couple of utility companies are currently testing this technology, there is only one coal-burning power plant in the world that's been using this method

---

259 Ibid.
260 "Mara Hvistendahl, "Coal Ash Is More Radioactive than Nuclear Waste," *Scientific American,* December 13, 2007.

successfully for years: the Basin Electric Power Co-Operative in North Dakota. The plant was built in the 1980s with the goal to convert coal into natural gas, not to fight global warming. To perform this transformation, carbon dioxide must be removed from the coal through the liquefaction process. Basin Electric does this efficiently and then stores the extract in former mine shafts and caverns beneath the ground.

Environmentalists appease the coal industry by feigning their approval of carbon sequestration, and Chu is likely among that crowd, as this statement he made to the World Affairs Council would indicate, "... we don't know enough about this technology, and that, by the way, ignores the legal resistance, 'not in my back yard' cost . . . the NIMBY [Not In My Back Yard] guys will go after carbon sequestration as well."[261]

Translation: Should sequestration technology be employed, his NIMBY environmentalist friends will scream like snotty toddlers over the liquefied carbon being pumped beneath the earth's crust.

Sadly, King Coal does not figure into America's energy future under the Obama administration. With the plans being concocted for cap-and-trade, power derived from coal will become more costly and less available—as will another reliable staple in the United State's energy portfolio: natural gas.

NATURAL GAS OFF LIMITS

In March of 2008, sectors of the media were abuzz about an interview Sierra Club leader Carl Pope gave to *Oil and Gas Investor Magazine*. Some heralded the interview as a sign that the environmental movement had finally mellowed its opposition to a fossil fuel, in this case natural gas—a product that provides about a quarter of America's total energy supply. But that's hardly what Pope said. Read the interview carefully:

> The Sierra Club is so opposed to coal as a fuel source that it has come out in support of U.S. natural gas, according to Pope's comments. It still

---

261 Stephen Chu, "A New Energy Program," speech presented to the World Affairs Council, San Francisco, September 13, 2007.

favors renewable fuels—solar and wind—more than fossil fuels, but among fossil fuels, it favors natural gas, Pope says.

"Among the fossil fuels, natural gas is at the top," says Pope, who became executive director of the roughly 700,000-member organization in 1992 and has seen some 150,000 members added during that time. Pope has been on the Sierra Club staff for some 30 years.

In second place, by the Sierra Club's score, among fossil fuels: crude oil. In last place: coal. [262]

Pope's statement is hardly an endorsement of the natural gas industry. Gauging from the many arguments I've conducted on my radio show with extremists who are members of Pope's club, their leader was simply implying he doesn't hate natural gas as much as he hates coal and oil. It would be like asking a conservative, "Marx, Lenin, or Stalin: who do you hate least?"

Natural gas is an abundant resource, free of significant pollution. It also produces about 30% less carbon dioxide than gasoline, and about 45% less than coal. But it still exudes CO_2, and as a result, remains squarely on whiners' hate list.

There are quadrillions of cubic feet of natural gas in the United States with more reserves discovered each year. It's estimated that at the current supply and rates of consumption, we have enough natural gas to last 118 years.[263] Besides the carbon dioxide flap, the problem are the NIMBYs who don't want nature to be disturbed during the extraction of this abundant fossil fuel, and once again, they're clearly winning the battle.

During the Clinton-Gore administration, millions of square acres of federal land, brimming with natural gas, were purposely placed off limits from the energy industry. Combining with state legislation, laws were also created to prevent 85% of the natural gas offshore from being extracted, including gas found along the coasts of California, Washington, Oregon, Alabama, Mississippi, Florida, Georgia, the Carolinas, Virginia, Maryland, New Jersey, New York, and New England.

---

262 "The Sierra Club Comes Out in Favor of the U.S. Natural Gas Industry, Reports Oil and Gas Investor Magazine," *PRNewswire*, March 8, 2008.
263 North American Natural Gas Assessment, Prepared for American Clean Skies Foundation by Navigant Consulting, Inc, Chicago, Illinois, July 4, 2008.

Despite its efficiency and availability, natural gas is simply not seen by those calling the shots in Washington as a future resource we should count on. Therefore, if a clean fossil fuel like natural gas is censored, and coal is cursed, is it any wonder that the primary fossil fuel which we all depend upon is also damned?

FOR SPACIOUS SKIES

For years we've been lectured that the United States has an insatiable lust for oil. We consume more of the black gold than any other country, we're scolded; with the lion's share refined into gasoline, which is used to operate our oversized cars, monster trucks, enormous SUVs, and luxurious airplanes. I've never been bothered by our oil consumption—we're the world's most productive country, and, despite our penchant for petroleum, we've done a remarkable job cleaning up our air.

Statistics compiled by the American Enterprise Institute reveal that in the 25 years spanning 1980 and 2005:[264]

- Fine particulate matter declined 40%.
- Ozone levels declined 20%, and days per year exceeding the 8-hour ozone standard fell 79%.
- Nitrogen dioxide levels decreased 37%, sulfur dioxide dropped 63%, and carbon monoxide concentrations were reduced by 74%.
- Lead levels were lowered by 96%.

What makes these air quality improvements even more noteworthy is that they occurred during a period in which:

- Automobile miles driven each year nearly doubled to 93% and diesel truck miles more than doubled to 112%.
- Tons of coal burned for electricity production increased over 60%.

264 Joel Schwartz, "Facts Not Fear on Air Pollution," NCPA Policy Report #294, December 2006, ISBN #1-56808-167-7.

- The real dollar value of goods and services (gross domestic product) more than doubled to 114%.

Despite more people and more cars on the road than ever, air pollution is no longer a growing issue in the United States; thus, the new bogeyman in that of CO_2. And, while the clean-up of our skies has gone underreported, so has our domestic supply of potential oil.

BLACK GOLD

No one is sure exactly how vast the earth's oil reserves actually are, as each year incredibly large pools of untapped crude are discovered, but we do know that currently the world's nearly 7 billion residents consume 86 million barrels of oil per day. As of January 1, 2008, proven, or recoverable, world oil reserves (oil reserves tapped or waiting to be tapped) as reported by the *Oil & Gas Journal*, were estimated at 1,332 *billion* barrels—14 billion barrels (about 1%) higher than the estimate for 2007.[265] Using a classic scare strategy, environmentalists crunch these numbers, crying that, at current rates of consumption, the earth will be out of oil in 41 years. But that's simply *not* true. There is something the oil biz refers to as "other" or "undiscovered" reserves. These are known reserves that are essentially more difficult to tap.

In the United States, there are two undiscovered reserves that, if drilled, would make our nation totally oil independent. The first is the Bakken formation in North Dakota and Montana. In 2008, the United States Geological Survey announced that up to 4 billion barrels of recoverable oil were discovered in this region.[266]

The second find, described as the largest untapped source of oil in the world, is the Green River reserve, located deep below federally-owned

265 International Energy Outlook 2008, Energy Information Administration, Official Energy Statistics From The U.S. Government, Report #:DOE/EIA-0484(2008). Release Date: June 2008.
266 "3 to 4.3 Billion Barrels of Technically Recoverable Oil Assessed in North Dakota and Montana's Bakken Formation—25 Times More Than 1995 Estimate," United States Geological Survey press release, April 10, 2008, www.usgs.gov/newsroom/article_pf.asp?ID=1911.

land in Colorado, Utah, and Wyoming. Many geologists estimate there may be 1.8 *trillion* barrels of oil awaiting recovery.[267] Known as "shale oil," it is more costly to extract, but even if only half can be tapped, scientists believe the amount is nearly *triple* the oil reserves of Saudi Arabia, the world's number one oil exporter.

Keeping us from the energy independence such reserves promise are the predictable environmentalists. They've been effectively blubbering about the billions of barrels of easily recoverable oil that has been placed off limits in the Alaska National Wildlife Reserve (ANWR) for decades. Activists claim ANWR is too pristine to drill, but that's sheer nonsense. So remote is this northern land that it's pitch dark three months out of the year, covered with snow six months out of the year, and once the sun peers out and the snow finally melts, is covered with caribou poop and swarming with mosquitoes the size of small birds the remaining three months of the year. Like parts of Nevada, which are ideal for storing nuclear waste, parts of Alaska are meant to "drill, baby, drill," as the locals often advertise.

Likewise, our oil reserves in the Gulf of Mexico and off the California coast are the envy of the oil-producing world, but once again, the enviro-whackos and their legislative stooges have placed most of this black gold out of reach.

Tapping these immense domestic reserves would allow the United States to become truly self-sufficient, but the global whiners are not interested in oil independence. They are bent on putting an end to the oil industry, period. Just look at how they treat the oil refineries.

ARCHAIC REFINERIES

Refining oil involves the complex process of morphing thick, dark crude into materials that can be used to fuel transportation, lubricate machinery, and create a host of other useful products including plastics. Because of intense pressure from environmentalists and increased governmental

---

267 James T. Bartis, Tom LaTourrette, Lloyd Dixon, D.J. Peterson, Gary Cecchine, "Oil Shale Development in the United States," prepared for the National Energy Technology Laboratory, United States Department of Energy, by the Rand Corporation, 2005.

regulation, the United States has experienced a steep decline in refining capacity. In 1950 we had 324 refineries; today there are 149. Since 1976, obtaining a permit to build a modern refinery has been so difficult and costly that no new refineries have been built. The only way for the oil companies to keep up with demand has been to expand existing plants and run them 24/7, placing the consumer at risk. When a refinery experiences a mechanical problem, or, for example, when a hurricane hits the Gulf Coast (home to many refineries), the supply chain is immediately disrupted, resulting in higher prices for gasoline at the pump. Refineries also act as storage facilities, thus increasing the number of refineries would equal an increase in available supply, resulting in lower prices for us.

For years, the ridiculous assertion that the oil companies are in cahoots to make sure their combined refinery capacity is purposely low so that they can manipulate the price of oil has been floating around, but just the opposite is true. As publicly traded, competitive companies, they would all love to try and increase production because it would boost their market share and value to their shareholders, plus deliver fuel to consumers at a lower cost.

But instead of forcing the environmentalists to step aside and allow the oil companies to expand their refining capabilities, Washington bureaucrats have bizarrely devised their own absurd scheme to appease the activists by the production of a bogus alternative known as biofuel.

BIOFUEL BUST

On December 19, 2007, George W. Bush signed the Energy Independence and Security Act into law, mandating that 36 billion gallons of biofuel be produced in the United States every year by 2022, a nearly five-fold increase over 2006 production levels. Spotlighting that it was among the dumber moves President Bush had made, then-Senator Obama, "House Madam" Pelosi, and Senate Leader Harry Reid all enthusiastically voted for the Bush biofuel plan, and have since repeatedly pledged to increase biofuel production far beyond the Bush mandates.

The popular biofuels are ethanol (essentially vodka mixed with gasoline) and biodiesel (pretty much the same as straight-up cooking oil). Biofuel is primarily processed from food crops, which include corn, soybeans, rapeseed (canola oil), sugarcane, and palm trees (as in palm oil). Farmers who grow these crops for fuel instead of food are motivated by whopping government subsidies.

The United States is the world's corn king, and, thanks to the federal rewards for going bio, an outrageously disproportionate percentage of that crop is now ending up in gas tanks instead of tummies. For example, according to the United States Department of Agriculture, a bushel of corn (56 pounds) is required to make 2.7 gallons of ethanol.[268] Therefore, to fill an 18-gallon tank with Ethanol demands nearly 370 *pounds* of corn! The result has been a shortage of corn for food and a subsequent increase in food prices globally. South of our border, the situation is so severe that price of tortillas has risen 400%, causing tens of thousands of angry Mexicans to take to the streets in protest.[269]

Likewise, biodiesel production has impacted the cost of food in general. Besides the price of cooking oil sky-rocketing, even crops used for animal feed are now being diverted to diesel fuel production, thus negatively impacting prices in the feed sector. The chain reaction from the boneheaded decision to encourage farmers to go bio has forced the price of milk, eggs, cheese, and meat, through the roof.

And there's more. Production of biofuels requires large amounts of nitrogen fertilizers, which, in turn, has created a fertilizer shortage. The price of fertilizer rose more than 200% in 2007 alone. This explains why the cost of virtually all the produce in our grocery stores has risen so steeply in the last couple of years and why even the price of processed foods that include rice and wheat have also ballooned in price.

Adding to all the lunacy, ethanol requires more gasoline to *produce* than it actually provides. After the crops are harvested using gas-powered vehicles, the ethanol is processed and refined at gas-powered plants. Because ethanol contains tiny amounts of water, it cannot be pumped through pipelines to be mixed with gasoline at the oil refineries because H_2O will produce rust and corrosion. That means the ethanol has to be

---

268 Allen Baker, Steve Zahniser, "Ethanol Reshapes the Corn Market," *Amber Waves* magazine, United States Department of Agriculture Economic Research Service, April 2006.
269 "Mexican stage tortilla protest," *DDC News*, February 1, 2007, http://news.bbc.co.uk/2/hi/americas/6319093.stm.

trucked from its production facilities to refineries where it can be blended with gasoline—wasting even more fossil fuel.

Altogether, 70% more energy is required to produce ethanol than the energy that actually is in ethanol.[270]

SOLUTIONS

America desperately needs to begin planning for additional energy supplies. By 2050 census data projects that the population of the United States will have increased from 310 million to 404 million,[271] and sadly, as you're about to see in the next chapter, the only energy plan being executed by the federal government is one that involves severe restrictions and limitations on the amount of energy you will be allowed to use.

At some point in time the American people need to stand up to the environmentalist community and tell them to "back off." We need a real energy policy—one that will provide inexpensive and abundant power for both now and the future.

Given that CO_2 is *not* altering the climate, we should responsibly harvest our vast fossil fuel resources. New coal-fired power plants, fashioned with superior pollution reduction technology and capable of increased efficiency, should be constructed, especially in the coal mining states. Natural gas fields beneath federal lands should be leased to the highest bidder, with instructions given to ensure that the extraction footprint is as small as reasonably possible.

Likewise, our domestic oil reserves need to be tapped, thus separating the United States from the teats of foreign oil producers who do not share our ideological interests. And we must let it be known that by utilizing the technological advances developed on Alaska's North Slope, *only* 2% of the surface of an oil field needs to be developed. In addition to the

---

270 According to Dr. David Pimentel, Professor, College of Agriculture and Life Sciences, Cornell University, as cited in the September, 2001 issue of the Encyclopedia of Physical Sciences and Technology.

271 Steven Camorata, "The Impact of Immigration on U.S. Population Growth," prepared testimony before U.S. House of Representatives, Committee on Judiciary, August 2, 2001.

continental reserves, our vast offshore oil reserves need to be harnessed—the days of leaks and spills are long behind us.

Reaping these fossil fuel resources from federal lands would provide our government a steady flow of revenue and in all cases would provide for thousands of permanent jobs.

Of course, if the debate over carbon dioxide continues, then the best compromise should be the nuclear option. In June 2009, Tennessee Senator Lamar Alexander delivered a speech to Congress and stated, "I propose that from the years 2010 to 2030 we build 100 new nuclear reactors to match the ones we are already operating. That is what we did from 1970 to 1990. During that 20-year interval, we built almost every one of the 104 reactors that now provide us with 20% of our electricity. If we build another 100 by 2030, we will be able to provide well over 40% of our electricity from nuclear power."[272]

Talk about a stimulus: in addition to bountiful electricity—according to the Nuclear Energy Institute—building 100 new nuclear plants would generate up to 240,000 construction jobs. Once complete, each facility would employ 700 permanent employees, or a total of 70,000 jobs (such positions usually receive 36% more pay than the average jobs in the community). In addition to the salaries paid to its employees, the average nuclear plant generates approximately $430 million in annual sales of goods and services with other business in its community; multiply that by one hundred and we have just created a national stimulus of $43 billion per year.[273]

But these are optimistic plans emanating from a heart of liberty, and currently those calling the shots desire nothing of the sort.

272 Senator Lamar Alexander, (R-TN), "Build 100 nuclear plants in 20 years," presented on Senate floor, June 2, 2009.
273 "Nuclear Power Plants Contribute Significantly to State and Local Economies," Nuclear Energy Institute, Fact Sheet, January, 2009. http://www.nei.org/resourcesandstats /documentlibrary/reliableandaffordableenergy/factsheet/nuclearpowerplantcontributions/.

10

CASH AND CONTROL

. . . for the love of money is the root of all kinds of evil.

—Saint Paul the Apostle

IN THE SHADOW of Lady Liberty, seeds of greed and corruption sown over the last 40 years are beginning to bear much evil, Marxist fruit. Self-anointed elitists have positioned themselves to seize political and financial power in our country, transferring wealth into huge sums of money for themselves while conning the American people into believing that a ruling class are the only ones worthy and capable of managing society and its problems. Having aggressively and unapologetically appointed themselves as the arbiters of human happiness, they justify "for the common good" mechanisms of broad control, crushing everything they perceive as archaic, including our founding ideals of self-determination, independence, and patriotism. These elitists are the enemies of freedom, and their godless grand strategy is more than seditious—it's purely evil.

In his 1796 farewell address from public service, our nation's first president, George Washington, ominously warned:

> Of all the dispositions and habits which lead to political prosperity, religion and morality are indispensable supports. In vain would that man claim the tribute of patriotism who should labor to subvert these great pillars of human happiness—these firmest props of the duties of men and

citizens. The mere politician, equally with the pious man, ought to respect and to cherish them. . . . It is substantially true that virtue or morality is a necessary spring of popular government.

America now finds itself at a dangerously redefining moment in its history. Orwellian politicians parading as patriots have joined powerful forces with corporate businesses claiming to champion the free market, but in reality, their anticapitalism schemes are quickly dismantling the last truly free country in the world under the guise of saving the environment. If they manage to succeed, life, liberty, and the pursuit of happiness as we have known it for the last 230-plus years will be trampled and strewn on the ash heap of history under malevolent and absolute tyrannical power.

MY SILICON VALLEY MOLE

In Chapter 5 I briefly introduced you to a man with whom I was acquainted through a charity to which we both contributed. For credibility's sake, allow me to share a bit more about this interesting fellow, who has since become a good friend. "Dave" we'll call him, is a first-generation American who dearly loves this country. He played college football, received a degree in computer engineering, joined the military, became a member of one of our most highly respected special forces ("I could basically kill anyone with my bare hands," he once told me with a hearty laugh), wrote a critical code that virtually everyone uses on their computer, started some successful companies, manages a hedge fund, is associated with several powerful venture capital firms in the Silicon Valley, and is a very generous guy who uses his money for good works.

Germane to the premise of this book, Dave fully grasps what is about to happen to his country, my country, our country.

"The coming energy shortage is about controlling our lives, and a host of complicit players, who believe such control is necessary and appropriate, are working with the government to assure this," he told me over coffee in Cupertino, just a few blocks from Apple's headquarters in the heart of the Silicon Valley. "These players are big fish from the financial world and from here in the Valley: Goldman Sachs, Kleiner Perkins, Google, GE, Microsoft, IBM."

"What are their motives?" I asked.

"It's two-fold. For some, it's a pure profit play. They're privy to the analyses that don't make it into the *Wall Street Journal*. The global economy is teetering on the brink of collapse, nearly every country on the planet is over-leveraged, and these money guys are looking at perhaps the last big opportunity of their lifetime."

"Like a 'green-bubble,' instead of the Internet bubble of the '90s," I interjected.

"Exactly. Private entrepreneurs with little more than a slick Power Point presentation, a star-studded board of directors, and an unproven idea for saving the environment will be funded by the venture guys. The valuations of the newly-funded companies will increase; eventually they'll get purchased by a bigger company, and the venture guys will get paid back handsomely. A few years down the line, the bigger company will either discover the business they bought is worthless, or they'll just fold and go under—like 80% of the Internet companies did. It's a bubble where jobs will be temporarily created, markets will briefly pick up, and the guys on Sand Hill Road [where many of the Silicon Valley's venture firms are located] will score."

"But the government's plans for things like the SmartGrid, keeping us from drilling for our own oil and natural gas, stopping nuclear energy, cap-and-trade—how can these money guys be so willing to let America go to hell?"

"I've heard you talk about Marx's Laws of Matter on the radio—these people *are* elitists. They believe Marx had some good ideas."

"What about global warming, are they believers?"

"They're just like Al Gore—they believe in money."

Our conversation concluded with Dave reaching into his satchel and retrieving a stack of documents two inches thick.

"I've heard your theories on the SmartGrid. They're spot on. The government will be able to control every electrified component of your house. Now," he said, handing me the stack, "here's some paperwork on the money trail associated with the grid, plus who's going to become rich from cap-and-trade. There's a lot here about my 'pal' Al, too. I'm not giving you anything that's classified; it's all pretty much open source. This'll just save you a couple years worth of research," he exclaimed with that hearty laugh.

I gratefully received the documents, collected my notes, and stood to shake hands with this great American. I couldn't wait to get home to begin to read his materials. His valuable input would be the capstone to an investigation I had begun over a year earlier.

IT STARTS WITH YOUR THERMOSTAT

During the first week of January in 2008, a caller to my radio program alerted me to proposed revisions in the gargantuan, 230-page California Building Code. The changes, he said, were being stealthily concocted by the California Energy Commission, and mandated that remotely controlled home thermostats be installed in all new or remodeled homes or in existing homes in which the furnace or air conditioner was being replaced. The public had 30 days to respond.

Conducting an immediate investigation, I learned the caller was correct. The plan involved a new technology known as the Programmable Communication Thermostat, or PCT. Embedded within the PCT is an FM radio receiver, which would allow energy authorities to control home temperatures. Customers would be denied the ability to override settings. Officials at the Capitol told me the PCTs would help avoid the many "rolling blackouts" we often experience during periods of peak power demand in California.

Because there are not enough electrical generation facilities in California (the environmentalists protest virtually every new planned power plant—and usually win), on hot summer days, when energy demand soars, certain sectors of the state suddenly see their power temporarily shut off. After a few minutes, the power returns, and another section goes out. This obnoxiously interrupting procedure continues until the power shortage abates. Such rationing techniques are daily experiences in third world nations.

With PCT technology, a thermostat czar would regulate the settings of every air-conditioned home in the state, thereby reducing demand on the grid. If there were ever a natural gas shortage, the same technique could be employed to dial-down furnaces.

Usually disagreements regarding the thermostat setting in my home are resolved between my wife and me, when—man that I am—I *allow* her to win. However, with these plans from the energy commission, some anonymous bureaucrat would take the decision of our indoor temperature away from my wife and me however they saw fit.

I went into battle mode—something I'd learned well working at KSFO.

In December 2003, I was on the air with now-nationally syndicated talk show host Melanie Morgan, and Shawn Steel, then chairman of the California Republican Party. We were discussing the poor performance of our governor, Gray Davis, when I naïvely asked, "Is there a mechanism for recalling the governor?"

Melanie grabbed that bull by the horns and rode it all the way to seeing Davis get a boot in the butt from the voters in a special election ten months later, opening the door for Arnold Schwarzenegger to try his hand as governor.

Remembering the Davis recall from five years earlier, I proudly borrowed a page from Melanie's playbook and shared the details of the thermostat plan with my audience, who, in turn, flooded the energy commission with phone calls, emails, voicemails, letters and protests. Within two weeks, the commission backed off their plans, but, as one staff member privately assured me, they'd eventually find a way to get this one done. They would have to—it was "part of a much larger plan."

INTRODUCING "SMART" TECHNOLOGY

My curiosity triggered, I began to research the "larger plan." The PCT was part of an emerging technology known as "demand response technology" or "direct load control." The technology allows the utility company to charge customers for electricity based on *time of day* usage and ultimately provides the company with the ability to turn down, or even interrupt, the power flowing into the customer's home.

Cleverly dubbed "smart" technology, the intention is to eventually ensure that all major appliances, including washers, driers, dishwashers, refrigerators, water heaters, and even lighting, can be governed remotely. Understand that this isn't some futuristic theory from the Jetsons. The time of George and Jane Jetson is upon us, but with far less societal benefits that the old cartoon ever promised. In 2009, General Electric (GE) received an award from the U.S. Department of Energy and the Environmental Protection Agency, as the federal government's Energy Star Partner of the Year for their "advances in 'smart' or Management Enabled Appliances designed to modulate power consumption based on usage time and rates."[274]

"Management Enabled" is trade-speak for appliances that can be managed remotely by someone other than the owner. GE currently has a pilot program operating in Louisville, Kentucky, to determine the effectiveness of their "smart" appliances. However, unlike the reality to come, customers in the pilot program have the ability to override automated changes to the appliance's performance at any time.

Apparently, this is part of a global effort. Darrel Bracken, a regional president of GE competitor, Whirlpool, recently announced that by 2015, their company will "make all the electronically controlled appliances it produces—everywhere in the world—capable of receiving and responding to signals from smart grids."[275]

If population projections hold true and, as we discussed in the previous chapter, there are no plans to proportionately increase the amount of available energy to satisfy the needs of the consumer at present levels, then the populace will be forced to use less electricity. Smart technology would be Big Brother's answer to the coming, planned energy shortage: the populace—forced to use less energy—will be required by the government to adapt to the new energy-saving devices, which, in turn, will guarantee policymakers more control, while allowing select vendors the opportunity to become incredibly wealthy beyond imagination.

Rationalizing this technology for federal government officials, a popular "smart" consultant has written,

> Under certain predetermined peak periods, the utility interrupts or cycles the appliance to achieve its system goal of reducing peak usage. . . .

---

274 http://www.goodcleantech.com/2009/03/ge_picks_up_top_energy_star_aw.php.
275 "Smart Grid: New Products Promoting Wise Use of Energy," electric energy online, Larry Gill, May 11, 2009, http://www.electricenergyonline.com/?page=show_news&id=111197.

In most cases customers do not "notice" or suffer adverse consequences for the interruption or cycling[276] [quotations used by the consultant].

Once again, the elitist policymakers assume American citizens are so ignorant that we won't "notice" that the clothes in the dryer never got dry.

ENERGY POLICY ACT OF 2005

When the California Energy Commission attempted to implement their PCT plan, they were actually marching to soft orders derived from the Federal Energy Act of 2005, signed into law by George W. Bush. Section 1252 of the Act states, "It is the policy of the United States that . . . demand response [technology] . . . shall be encouraged."[277, 278]

In other words, the 2005 legislation was conditioning the public for future erosions of liberty, getting the salesman's proverbial foot in the door, so to speak, for the deal to be closed at a later date. But in discussing the ways of Big Government, perhaps the Trojan Horse analogy might be more accurate. Planning for the vigilant citizens who would surely oppose such intrusive metering devices, clever politicians allowed the Secretary of Energy the authority to educate "consumers on the availability, advantages, and benefits of advanced metering and communications technologies," and to work "with States, utilities, other energy providers and advanced metering and communications experts to identify and address barriers to the adoption of demand response programs."[279]

No surprise, then, that soon after concerned Californians stopped the energy commission's plan, California's major utility provider, PG&E, rolled out their own version of the PCT, called "SmartAC." This device would not replace the thermostat, but would, instead, be located outdoors, adjacent the customer's air conditioner. According to PG&E's marketing brochure:

---

276 "Smart Meters, Real-Time Pricing, And Demand Response Programs," a report prepared under contract for the Oakridge National Laboratory by Barbara Alexander, Consumer Affairs Consultant, May 30, 2007.
277 Ibid.
278 Addendum to Energy Policy Act of 2005, H.R.6-370, Section 1252 "Smart Metering."
279 Ibid.

When activated, your AC will do what it would normally do for about 15 minutes of every half hour—make cool air. For the other 15 minutes, your system fan will circulate already cool air and your AC will make cool air again when the next 15-minute cycle begins. You'll likely not even notice when this happens. In fact, in a survey of sample participants, most didn't notice a change.[280]

Like the federal consultant, PG&E makes the usurping of personal liberty sound so simple . . . "*most didn't notice a change.*"

SMART METERS

The remote controlled thermostat was just the first step in the smart technology scheme. But the next big step was the implementation of the SmartMeter, also subtly introduced in the 2005 legislation. Again, from Section 1252:

> Not later than 18 months after the date of enactment of this paragraph, each electric utility shall offer each of its customer classes . . . a time-based rate schedule under which the rate charged by the electric utility varies during different time periods. . . ."[281]

The SmartMeter replaces the old, spinning wheel, electric meter mounted on the side of your house, which is read monthly by a dog-weary utility company employee. Chances are likely that your old meter has already been swapped out. If so, your energy provider is already able to measure how much power you are consuming minute-by-minute (utilities are employing the same technology for natural gas and water meters). Despite the benevolent marketing spiel utility companies may be using to promote the benefits of smart technology, let me reiterate,

---

280 "SmartAC Details and Program Requirements," PGE.com, http://www.pge.com /myhome/saveenergymoney/energysavingprograms/smartac/programdetailsandrequire ments/index.shtml.
281 Addendum to Energy Policy Act of 2005, H.R.6-370, Section 1252 "Smart Metering."

this is *not* a helpful plan to assist you in meeting your family budget; it's a social engineering scheme, designed and promoted by the federal government to change your behavior. By having the capability to read your meter many times a day, the utility company is able to effectively establish demand pricing schedules; thus, using too much electricity during the peak periods is going to cost you.

And, probably like yours, my utility company has been pretty slick in selling this plan. From their website:

> PG&E is dedicated to providing you with the best service possible. We're always looking for new and innovative ways to make your daily interactions with us faster, easier and more convenient.
>
> That's why we're introducing the PG&E SmartMeter program. The SmartMeter program . . . will allow us to collect meter data without setting foot on your property and interrupting your schedule. Once the SmartMeter system is up and running in your area, we will collect energy usage data much more frequently—hourly for residential electric and daily for gas.[282]

Wow, how wonderful! They're concerned about *my* convenience and *my* privacy; and from now on they'll "collect meter data without setting foot" on *my* property, because they don't want to "interrupt" my busy schedule. My, my, my, how these people really care about me . . . *not*.

The SmartMeter, which eventually will be connected to every home, apartment, and business in America, is simply another increment in the planned transformation of the U.S. from a constitutional republic to a quasi-democratic dictatorship.

SMART GRID

The next phase involves the construction of the SmartGrid. The Grid was first brought to the mainstream's attention during the 2008 presidential

---

282 "Your meter is about to get an upgrade," Pacific Gas and Electric website, September 2008, http://www.pge.com/myhome/customerservice/meter/smartmeter/.

campaign, as Obama spoke of the "need to modernize our national utility grid."[283] And, like the PCTs and the Smart Meters, the Grid was also covertly introduced in the Energy Act of 2005. Section 1221 permits the construction of "a national interest transmission corridor designated by the [Energy] Secretary."[284]

This simple clause announced, in the cause of "national interest," the birth of the coming SmartGrid. Over the next 18 months following the bill's signing, as plans for the expansive corridor were being revealed, a spark of concern began to develop: the new transmission lines were going to tear through most states without their approval, violating the constitutional provision of state's rights. Eventually, token Congressional hearings were conducted. One of the more vocal opponents was Bill DeWeese, a Pennsylvania state representative who had learned that 57 of Pennsylvania's 60 counties had been declared a part of the federal corridor.[285]

DeWeese spoke before Congress, rightly arguing that if the Federal Energy Regulatory Commission "is permitted to use its congressionally conveyed authority to commandeer and usurp the traditional role of states and their administrative agencies to review and approve the location and construction of high voltage transmission lines, Pennsylvania, not unlike every other state, would have no control, no say, and no recourse other than expensive litigation; over transmission planning, location, and construction within its geographic borders."[286]

And DeWeese was correct—states were shut out of the discussions. Huge swaths of real estate would be commandeered for the feds' corridor of transmission towers. Only national parks, federal reserves, and regions protected under United Nations' biosphere agreements, would be spared.

---

283 Barack Obama's campaign speech, Lansing, Michigan, August 4, 2008.

284 Energy Policy Act of 2005, Subtitle B—Transmission Infrastructure Modernization, Section 1221, Siting of Interstate Electric transmission Facilities, amended to include Section 216, p. 354.

285 "DeWeese criticizes DOE decision on power line," press release, Pennsylvania Majority Leader Bill DeWeese, October 2, 2007.

286 "DeWeese seeks repeal of federal Energy Act," press release, Pennsylvania Majority Leader Bill DeWeese, May 11, 2007.

It wasn't until 2007 that the mysterious federal corridor was finally christened. The SmartGrid was outed by name in the Energy Independence Security Act and signed into law December 19, 2007. Buried in its 310 pages, an entire section was devoted to the coming grid. Some 3,000 miles of electric lines would be strung, a gigantic step toward updating the entire nation's existing infrastructure. However, proving that most members of Congress (and perhaps even the president) didn't know what they had agreed to-probably because they never read the bill, or, if they did, didn't bother to ask any questions—on December twentieth, the day *after* the legislation became law, a memo from the Congressional Research Service was distributed to members of Congress, describing how the SmartGrid works:

> The term SmartGrid refers to a distribution system that allows for flow of information from a customer's meter in two directions: both inside the house to thermostats and appliances and other devices, and back to the utility. It is expected that grid reliability will increase as additional information from the distribution system is available to utility operators. This will allow for better planning and operations during peak demand. For example, new technologies such as a Programmable Communicating Thermostat (PCT) could connect with a customer's meter through a Home Area Network allowing the utility to change the settings on the thermostat based on load or other factors. PCTs are not commercially available, but are expected to be available within a year.[287]

I'm sure the memo was recited like a cheap script to the thousands of concerned Americans who *had* read the bill and were phoning their congressman's office to complain about its passage. The fact that such communiqués are not issued *before* legislation is voted on—let alone after signed into law by the president—illustrates the incompetence within our nation's elected leaders.

As mentioned in the congressional brief, besides the transmission towers and new power lines, the SmartGrid will include the PCT, the

---

287 Amy Abel, Specialist in Energy Policy, "Smart Grid Provisions in H.R. 6, 100th Congress," Congressional Research Service, December 20, 2007.

SmartMeter, and a key software package known as the Home Area Network (HAN). The HAN will interface with all Energy Star appliances and other electrical devices within your home, including your TV, computer, and eventually lights, and will be connected to the Internet via a dedicated broadband connection.

Once the PCT, HAN, and SmartMeter are connected to the SmartGrid, Big Brother will become the new authority figure in your home. With a bureaucratic keystroke, anything attached to your Home Area Network could be selectively turned off—or on—without your approval.

Originally, according to Section 1306 of the Energy Independence Security Act, funding the SmartGrid was to be shared by the various states. However, this funding mechanism posed a serious problem. Most states were working with rapidly thinning budgets, and it was unlikely that they would be willing to invest the billions necessary to upgrade to the new grid. Likewise, most utility companies were unable to invest the necessary capital for improvements.

The dilemma was rectified in the 2009 "stimulus" bill, which authorized $16.8 billion in direct spending by the U.S. Department of Energy's Office of Energy Efficiency and another $4.5 billion to upgrade the nation's grid, plus at least $2.8 billion for installing broadband—another component critical to the success of the SmartGrid. Thus, with the first stroke of his golden presidential pen, Obama lifted the burden of funding the Orwellian SmartGrid off the slumping shoulders of the states, and onto the backs of the American taxpayers. Those involved in creating the grid's components would be guaranteed a future payday.

And eventual implementation of the grid would be easy. The PCTs, SmartMeter, and HAN will be offered for free by the utility companies eager to boost their revenues. New Energy Star appliances will be marketed with government rebates (as they already often are). And for those citizens too stubborn to adapt, "Ve haf vays of making you comply"— building codes simply will be altered to *force* compliance. Nestled in the 2009 America Clean Energy and Security Act are federally mandated energy-efficient building regulations, which supersede *all* local and state codes and which will be enforced by a national, green goon squad, funded in part by revenues from energy taxes, as well as by an annual $25 million from the Department of Energy "to provide necessary enforcement of a national energy efficiency building code. . . ."[288]

The legislation would also authorize the Secretary of Energy to "enhance compliance by conducting training and education of builders and other professionals in the jurisdiction concerning the national energy efficiency building code."[289]

Thus, if enacted, each time a home is built, remodeled, or prepared to be sold, a G-man wearing a federal badge and armed with a clipboard will show up at your house to make sure you have a PCT, that all of your appliances have been updated with the most recent Energy Star–approved internal communication devices, and that the Home Area Network has been properly installed and connected to your new SmartMeter, whether you like it or not.

FOLLOW THE MONEY

When Saint Paul spoke of "the love of money" being the root of all kinds of evil, he wasn't saying *money* is evil—money is a great tool for meeting needs, providing security, generating charity, and creating happiness. But *the love* of money is different—its pursuit can justify a host of grave deeds.

As I sifted through the details Dave provided, my convictions were confirmed; many of those fiercely peddling anthropogenic global warming are consumed by the *love* of money and stand to profit greatly from the perpetration of this fraud.

I believe such motives are currently driving General Electric (GE). GE is a major manufacturer of SmartMeters, the leading global manufacturer of wind turbines, and one of the nation's premier appliance manufacturers. They also have an influential hand in solar panel production and are the primary producer of those obnoxious, toxic, curlicue compact fluorescent light bulbs. The global warming scam continues, and will continue, to be the best thing that's happened to GE since its founder, Thomas Edison, invented the incandescent light bulb that revolutionized the entire world; an incredible irony, given his invention has since been

---

288 America Clean Energy and Security Act of 2009, page 323, June 22, 2009.
289 Ibid., p. 319.

vilified and, thanks to the federal 2007 Energy Independence Act, is set to be outlawed by 2020.[290]

Cries of climate change are bailing out GE's otherwise lackluster business enterprise, and their lobbying efforts clearly demonstrate their dependence on this environmental hoax for survival. In the fourth quarter of 2008, as the company's stock fell 30%, GE spent $4.26 million lobbying Congress to get their share of the then-forthcoming stimulus aid and global warming legislation anticipated to be signed into law under incoming President Obama—that equals $46,304 of daily lobbying, including weekends and holidays.[291] All told, GE's 2008 lobbying bill amounted to $18.66 million.[292] And, NBC, GE's television network, has devoted millions of dollars of free air-time to further the global warming deception.

You may recall that in 2007 the network ran a week of "green" programming to raise awareness about climate change. NBC kicked off the coverage during a Sunday night football game between the Dallas Cowboys and the Philadelphia Eagles. During the halftime show, studio lights were dimmed to save energy, while anchor Bob Costas looked like a blithering idiot reporting from the semidark set, his face ridiculously illuminated by candles. Meanwhile, a gigantic 72-inch, high-resolution, energy-sapping television monitor, provided a warm glow over his shoulder, and two 42-inch screens advertising Toyota hybrids burned brightly in front of his announcer's desk.

During the show, Costas segued to *Today Show* host Matt Lauer, who had been flown to the Arctic Circle with a full TV crew. And there he stood in the middle of the Arctic night, snow stretching for miles in every direction, illuminated by several thousand kilowatts of dazzling artificial light, making up a hundred times over for the symbolic dimness back at the studio, not to mention, of course, the tankfuls of fossil fuel burned to transport the entire crew to such a remote location. Anyone with a brain should have been repulsed by NBC's embarrassingly blatant hypocrisy.

---

290 2007 Energy Independence Act, H.R. 6, at 89.
291 "Obama's hidden bailout of General Electric," *Washington Examiner*, Timothy P. Carney, March 3, 2009, http://www.washingtonexaminer.com/politics/Obamas-hidden-bailout-of-General-Electric_03_04-40686707.html.
292 Ibid.

Without the boost they're getting from the global warming scare, GE would likely go the way of Westinghouse—into the tomb of once-great American companies that crashed and burned because of poor management. But give GE politically-correct credit: their CEO, Jeff Immelt, played his crony cards perfectly. He was a big supporter of Obama and consequently was awarded with a seat on the president's White House Economic Recovery Advisory Board.

GE's payday was announced in June 2009, when Obama and Energy Secretary Chu held a joint press conference revealing their plan to "expand and accelerate the development, deployment, and use of energy-efficient technologies in all types of commercial building as well as new and existing homes."[293] It was a sign of the green bubble to come, with GE set to experience a phenomenal business revival, supplemented with federal subsidies for research and development.

SMARTGRIDDERS: GOOGLE, MICROSOFT, AND IBM

Within Dave's stack of stuff were documents revealing the significant stakes technology giants Google, Microsoft, and IBM have in developing components critical to the completion of the national SmartGrid. My own personal research indicated that personnel at these three companies were also financially in bed with Obama, and, in some cases, advisors to his campaign and presidency.

According to the Center for Responsive Politics, Google employees and family members contributed more than $790,000 to Obama's campaign, Microsoft's people gave more than $806,000, and those working for IBM contributed more than $518,000.[294] In contrast, Republican presidential nominee John McCain received about $20,000 from Google employees, $63,000 from Microsoft's, and $52,000 from IBM's.

---

293 "Obama Administration Launches New Energy Efficiency Efforts," Office of the Press Secretary, The White House, June 29, 2009.
294 Center for Responsive Politics, Barack Obama Top Contributors, Based on Federal Election Commission data released electronically on Monday, March 02, 2009. http://www.opensecrets.org/pres08/contrib.php?cycle=2008&cid=N00009638.

Because of his business interests, Dave has spent considerable time on the Google campus. According to his description of the search engine's corporate culture, their belief in anthropogenic global warming is almost cult-like—and for obvious reasons. For starters, Al Gore is on their advisory board. Dan Reicher, Google's Director of Climate Change (they actually have such a division) reportedly assisted with the Cleantech for Obama committee, which raised about $2 million in campaign contributions.[295] Reicher was also a member of the Obama transition team and is thought to have been on the shortlist for the energy secretary position. Google's CEO, Eric Schmidt, who also vigorously campaigned for Obama, has attended White House economic summits and reportedly was considered for the administration's technology czar.

Hoping to cash in on its political associations, Google has developed a software application called the PowerMeter, which tracks home energy usage online. Dave believes the software could become the foundation for the Home Area Network. Because the product requires the appropriate hard infrastructure, Google has partnered with GE in an effort to get a cut of the sales of millions of SmartMeters.

Not to be outdone in the race for the HAN, Microsoft has developed its own online product known as "Hohm." It, too, would integrate with the SmartMeter, giving Microsoft a wedge into this lucrative market. Peddling their influence, Microsoft CEO Steve Ballmer, founder Bill Gates, and their spouses each gave $50,000 to Barack Obama's inauguration.[296]

IBM is also jostling for a piece of the action, as evident by their vice president of energy and utilities, Allan Schurr's, admission to *Politico*, "We've been speaking with the Obama administration quite a bit."[297]

In fact, according to the *Wall Street Journal*, well before the inauguration, during a conference call with Obama's transition team, IBM CEO Samuel Palmisano shared calculations suggesting that $10 billion invested in smart grid technology would create 239,000 jobs.[298]

---

295 "Obama says yes to Smart Grid," *Politico*, Erika Lovley, March 4, 2009.
296 Center for Responsive Politics, Barack Obama Inauguration Donors, http://www.open secrets.org/pres08/inaug.php?cycle=2008.
297 Ibid.
298 William M. Bulkeley, "IBM Chief: IT Investment Will Create Jobs," *Wall Street Journal*, January 6, 2009.

No doubt, jobs will be temporarily created by the build-out of the SmartGrid and its components, but how many other jobs will be permanently lost as a result of the other plans in the works?

CAPPED AND PLAYED

The rush to build and implement the SmartGrid is timed to coincide with the coming, *planned* mass energy shortage known as cap-and-trade. Cap-and-trade will eventually eliminate America's energy suppliers, dismantle our manufacturing base, increase every citizen's day-to-day living expenses, and be used as the honey pot for a host of welfare projects. It will enable the elitists to accomplish their goal to "de-develop" the United States. [299]

Every oil refinery, natural gas producer, electric utility company—any entity involved in the production of energy—will be limited by the EPA as to their allowable amount of CO_2 emissions. That's the "cap," and the cap will decrease each year. Likewise, other major carbon-emitting industries, such as steel, cement, glass, paper, lumber, mining, welding, airlines, trucking, and manufacturing, will also be hit with a cap based on an annual estimate of the tons of carbon dioxide each business releases into the atmosphere. Businesses will be assigned carbon "permits" corresponding to EPA-determined standards. One can only imagine the tremendous opportunity for graft, corruption, and favoritism carbon permits will allow and encourage.

If it's determined that any company in these carbon-emitting industries has exceeded their annual carbon permit quota, they may purchase "emission credits" from a government-approved exchange to offset their CO_2 emissions—that's the "trade." The carbon credits will likely be generated from two primary sources. Businesses able to maintain their emissions below the allotted permit cap will have surplus credits, and thus, will be able to trade those credits on the "carbon exchange." Credits, or "offsets," will also be awarded to major landowners who create or maintain carbon dioxide sinks by planting trees, not cutting down standing timber, or not plowing their fields. Those credits can also be traded on the exchange.

---

299 From this book's Foreword.

Each year, the government will auction off new permits to carbon producers and re-award credits to the offsetters. With each transaction, the feds will receive a transaction fee. Those with leftover credits from the prior year will be able to hold them for a specified period of time, and then, depending on demand, sell them later for a greater profit. Long sales, short sales, speculation, loaning credits for cash—it will all be possible with cap-and-trade. According to the Congressional Budget Office, by 2015, the federal government will be hauling in at least $104 billion a year from cap and trade.[300]

The most obvious repercussion of our concerned government hard at work managing our lives will be the increased cost of doing business in America, and all of the new costs will be passed down the line to the consumer. Virtually everything will be more expensive; and, despite what liberal politicians will tell you, jobs will be lost. Rather than participate in this cap-and-trade bait-and-switch, many manufacturing businesses will move operations to developing and third world nations, which have always been exempt from the United Nations' greenhouse gas treaties and initiatives, boosting their economies. And recall, this too is an acceptable facet of the overall plan. In fact, to prepare for this massive outsourcing, on page 1157 of the Clean Energy and Security Act we discover that for workers who lose their manufacturing jobs because the caps on their companies are too repressive, the "adversely affected worker" shall receive 70% of their prior weekly wage, "payable for a period not longer than 156 weeks." In addition, on pages 1170-1172 we read the unemployed worker can submit up to $1,500 in job search reimbursements, and get another $1,500 to cover his moving expenses. That's three years of unemployment benefits and three grand in perks!

However, while America is getting hammered with more expensive, less available energy, and, according to estimates from the Heritage Foundation, losing three million jobs shipped overseas,[301] greedy, green

300 Congressional Budget Office Cost Estimate, H.R. 2454, American Clean Energy and Security Act 2009, June 5, 2009.

301 "The Economic Impact of Cap and Trade," testimony by David Kreutzer, representing the Heritage Foundation, before the U.S. House of Representatives' Committee on Energy and Commerce, April 22, 2009. http://www.heritage.org/Research/EnergyandEnvironment/tst050709b.cfm.

fat cats will be socking away what could be their last big payday of their lifetime.

CARBON RICH

Unlike gold, wheat, or cattle, the new commodity—CO_2—is one you can't touch, taste, or smell. However, like the other commodities, a new exchange, much like the New York Stock Exchange, will be created. The job most likely will be awarded to a privately held company called the Chicago Climate Exchange (CCX). As mysterious as the inner workings of the Federal Reserve, CCX was supposedly created as the first voluntary cap-and-trade system established in the United States.

"CCX is an institution created with the sole purpose of being ready for the day when carbon trading in the United States will be the law of the land, complete with federal rules and regulations, much like the way carbon trading is conducted in Europe on the European Climate Exchange," Dave told me. "Betting on CCX to become the designated carbon trading depot is as easy as picking the sun to rise in the east tomorrow morning, especially since Obama and Gore have ties to the company."

It's true. Research reveals Obama and Gore *are* in on this. While living in Chicago, Obama served as a board member of the radical, nonprofit Joyce Foundation from 1994 to 2001. Joyce is perhaps the foremost advocate of trying to suck the individual right to bear arms from the Second Amendment. According to Ed Barnes of Fox News, in 2000 and 2001 "the Joyce Foundation gave nearly $1.1 million in two separate grants that were instrumental in developing and launching the privately-owned Chicago Climate Exchange."[302]

When the foundation made its first grant to the Climate Exchange, Joyce's president was Paula DiPerna. DiPerna left the organization in 2001 to become a founding executive vice president of CCX. In 2009, Barnes interviewed DiPerna and asked about Obama's role in the 2000

---

302 "Obama Years Ago Helped Fund Carbon Program He Is Not Pushing Through Congress," FoxNews.com. Ed Barnes, March 25, 2009. http://www.foxnews.com/politics /first100days/2009/03/25/obama-helped-fund-carbon-scheme/.

and 2001 grants. She replied that, as a director, Obama "read the proposal and voted on the grant."

And some of America's biggest moneymakers are lining up to get in on this new Chicago-based cash cow by skimming a cut right off the top.

"Every time a carbon credit is sold or purchased," Dave explained to me, "somebody will get a slice of the transaction—a fee. And, just like the way they trade in the commodities market, these smart investors will buy bulk-loads of credits on the cheap, and then position themselves to sell the credits to needy businesses for great profit. Carbon dioxide will be a new currency."

While perfectly legal because the law will allow it, as it always does when immorality is legislated, these greedy, conscienceless investors will rape and plunder our once-great nation with hearty political approval. And, of course, at the front of the line is Al Gore, in perfect position to cash in on his decades-worth of global warming hysteria. Dave agrees with reports that indicate the Chicago Climate Exchange is heavily influenced by Gore.[303]

"The guy is like Midas," Dave said of Gore, whom he has met in business meetings. "Just connect the dots on how this guy has made money in the last few years."

Dave produced a pen and paper, and began to draw an intricate web.

WEB OF FORTUNE

In 2001, following his failed presidential run, Gore, a man with no experience in the world of finance (unless you include "mistakenly" taking illegal campaign donations from a Buddhist Temple in California), co-founded the financial investment firm, Metropolitan West, with Philip Murphy, formerly of Goldman Sachs (GS) and previously the Democratic finance committee chairman.

---

303 According to Judy McLeod's story, "Obama's involvement in the Chicago Climate Exchange," *Canada Free Press,* May 25, 2009, Generation Investment Management "has considerable influence over the major carbon credit trading firms that currently exist, including the Chicago Climate Exchange." http://canadafreepress.com/index.php/article/9629.

Also in 2001, Gore was recruited to be an advisor to Google, picking up loads of stock prior to Google going public. Gore's stock is now worth untold millions.

In 2003, Gore's net worth shot even higher, as he took a position on the board of Apple, Inc., securing both an undisclosed director's salary and significant shares of stock. The same year, he sold Metropolitan West to Wachovia Bank for an estimated $50 million.

Around this time, former Metropolitan West partner Murphy put Gore together with David Blood, another Goldman Sachs partner, and the next year they founded Generation Investment Management (GIM) in London. GIM is a hedge fund that invests in companies supposedly committed to "going green." As of July 2008, GIM reportedly managed a portfolio worth $1 billion.[304]

Interestingly, another GIM founding partner is former Goldman CEO and U.S. Treasury secretary under George W. Bush, Hank Paulson. Other prominent former Goldman executives, including Mark Ferguson who ran European research, and Peter Harris, who oversaw international operations, were also founding GIM partners.

Though it cannot be confirmed through official financial statements (none are available), GIM is said to have a 20% stake in Europe's official carbon trading exchange and the European Climate Exchange (ECX), as well as a stake in the Chicago Exchange. Certainly many reputable publications, *Human Events* among them, have written about GIM appearing "to have considerable influence over the major carbon-credit trading firms that currently exist . . . including the Chicago Climate Exchange . . ."[305]

Gore's investment in both ECX and CCX would make perfect sense, given that public records indicate GS owns a large share of ECX, and bought a 10% stake in the Chicago Exchange. Indeed, Goldman's website boasts: "Goldman Sachs is active in the markets for carbon emissions. . . . Additionally, we have created new financial products to help our clients manage the risks posed by climate change. In September 2006, we made a

---

304 http://www.stockpickr.com/port/Al-Gore-Generation-Investment-Management/.
305 "The Money and Connections Behind Al Gore's Carbon Crusade," *Human Events*, Deborah Corey Barnes, October 3, 2000. http://www.humanevents.com/article.php?id =22663.

minority equity investment in Climate Exchange PLC, which owns several European and US trading platforms that facilitate trading in environmental financial instruments: the European Climate Exchange (ECX), [and] the Chicago Climate Exchange (CCX). . . ."[306]

In addition, the World Bank joined CCX in June of 2006, and now operates a Carbon Finance Unit that conducts research on how to effectively develop and trade carbon credits.[307] For Gore, cap-and-trade seems to be an easy means of cash and carry.

Another branch of the web involves Gore's extensive connections to the Silicon Valley. In 2007, Gore joined the foremost venture capital firm in the Bay Area, Kleiner Perkins (KP). KP is run by Gore's longtime pal, John Doerr. Among KP's big hits are financial backing for Google, Netscape, Sun, and Amazon. Upon joining the venture firm, Gore claimed that he would give 100% of his KP salary to his nonprofit Alliance for Climate Protection. But Gore's giveaway is quite possibly a total sham, because he's entitled to a tax write-off for the charitable contribution *and* is eligible to receive both a salary from the nonprofit, as well as paid expenses.

In 2008, Gore and Doerr started their "Green Growth" fund, hiring a team from Goldman Sachs' Special Situation's Group to hunt for their initial deals involving start-ups and young companies trying to make a play in the alternative energy field.[308] Dave believes the Green Growth fund, which is a mighty arm of the Kleiner Perkins portfolio, is worth close to a billion dollars, and also has an investment in CCX.[309] The fund also invested $75 million into Silver Spring Networks, a company that provides essential systems infrastructure for the SmartGrid.[310]

"Watch the companies tied to the Green Growth fund carefully," Dave

---

306 http://www2.goldmansachs.com/citizenship/environment/business-initiatives.html.
307 Karan Cappor, Phillippe Ambrosi, "State and Trends of the Carbon Market 2008," The World Bank, May 2008.
308 Alexander Haislip and Dan Primack, "Kleiner Perkins raising green growth fund," *Private Equity Week*, April 24, 2008, http://www.pewnews.com/story.asp?sectioncode=36&storycode=44384.
309 Confirmed by a story in the *New York Times*. "Gore's Duel Role: Advocate and Investor," John M. Broder, November 2, 2009.
310 "Kleiner Perkins Caufield & Beyers Lead $75 Million Investment in Silver Spring Networks, Smart Grid Technology Leader," *Business Wire*, October 7, 2008, http://findarticles.com/p/articles/mi_m0EIN/is_2008_Oct_7/ai_n29480769/.

told me, specifically noting the Silver Spring deal. "They'll score huge government grants, no doubt influenced by Gore's leverage on Capitol Hill, allowing the boys at Kleiner Perkins to recoup their initial investments many times over."[311]

2008 was also the same year Gore reportedly invested $35 million with Capricorn Investment Group LLC, a Silicon Valley firm that selects the private funds for clients and invests in makers of environmentally friendly products, according to a filing with the Securities and Exchange Commission.[312] Capricorn was founded by billionaire Jeffrey Skoll, former president of EBay Inc. and an executive producer of Gore's film. Dave also assumes Capricorn has an investment in CCX, but the firm is so secretive that even their website is password protected.

"The players at Capricorn are aggressive," Dave said. "Their goal is to make more money than anyone else in their space. Having Gore on their team is critical. "[313]

"My gosh," I said, staring at the web. The spirit of my old man was welling up within me. "This stinks. We're screwed."

"If you mean cap-and-trade is unstoppable—you may be correct. There are too many powerful people in this web."

"But it's not just cap-and-trade. There's the SmartGrid, the energy policies, and all of the disinformation on global warming—"

"You're right, Brian. It's massive. Our only hope is that enough Americans get educated real fast about these matters, and intellectually, politically, and prayerfully fight back. Short of voting these traitors out of office, if there's even time left to do that, it's going to take a divine miracle."

---

311 Confirmed by a story in the *New York Times*. "Gore's Duel Role: Advocate and Investor," John M. Broder, November 2, 2009. The article states, "... more than $560 million went to utilities with which Silver Spring has contracts, Kleiner Perkins and its partners, including Mr. Gore, could recoup their investment many times over in coming years."

312 Miles Weiss, "Gore Invests $35 Million For Hedge Fund With EBAY Billionaire," *Bloomberg*, March 6, 2008, http://www.bloomberg.com/apps/news?pid=20601070&sid =a7li9Nhmhvg0&refer=home.

313 Confirmed by a story in the *New York Times*: "Gore's Duel Role: Advocate and Investor," John M. Broder, November 2, 2009, Ion Yadigaroglu, co-founder of Capricorn stated, "We're trying to make more money than others doing the same thing and do it in a way that is superior...."

I sat there quietly staring at the web, almost stunned, trying to process the mix of emotions churning inside. Sensing my obvious despair, Dave smiled to encourage me, and himself, I think.

"Sussman, I was trained by the U.S. military to be a warrior. I won't stop trying to save this country, *and neither will you*," he soberly charged me.

I silently shook my head in agreement with my friend. He leaned forward across the table.

"That's why I want this book to be a huge seller."

And so, we—Dave, *you*, and me—*we* will never stop trying.

EPILOGUE

Facts are stubborn things; and whatever may be our wishes, our inclinations, or the dictates of our passions, they cannot alter the state of facts, and evidence.

—U.S. President John Adams

JANUARY, 2010. I'm sitting with my laptop, gazing westward through a large window, overlooking one of my favorite places on earth—Donner Lake, California. I've ventured here for the week to tap out the final words of this book. It's a cold but beautiful winter day. A few puffy cumulus clouds are randomly suspended in the brilliant blue sky, floating slightly above the snow capped 7,500-foot craggy peaks of the famed Donner Summit. It's difficult to imagine the courageous American settlers who forged those same unforgiving mountain tops 150 years ago, let alone the stranded pioneers of the Donner Party who perished during a series of ferocious storms on the south end of this very lake in 1846.

The last time I spent an extended stay here was during the week of July 4th. I recall looking out this same window, gaining inspiration, as I worked on this book. Scores of families were in view, relaxing on the nearby beach, enjoying life, liberty, and the pursuit of happiness. The weather was warm and the window glass was open, and I vividly remember the rhythm of the water lapping the shore while an American robin sweetly sang from the limb of the giant pine just outside. In the distance I recall the purr of powerboats towing wake-boarders who were enjoying the holiday weekend.

I wish you could have been here to witness the fireworks on the Fourth. It's an annual, spectacular, 30-minute display that beautifully lights up the entire beach and surrounding forest. Thousands of people pour into the area, carefully navigating the narrow road adjacent to the lake—the same trail that once led covered wagons from back east, up and over the summit and into California's expansive Central Valley. Our deck and dock were crowded with special friends and family, some having arrived by boat from the other side. Fellowship and patriotism overflowed. Everyone agreed that this year's fireworks show was the best *ever*—but, of course, just like you, we say that *every* year.

God, I love this country.

GLOBAL COOLING

I recall speaking with my mother on the phone over the Fourth. She lives in Chicago and told me it was the coldest Independence Day there in 30 years. I remember her saying Uncle Doug up in northern Minnesota had only planted his garden two weeks earlier—which was about a *month* late. Funny, Doug recently told me his tomatoes never really did turn out—the summer was just too darned cold—for the *third year* in a row. Doug is a great outdoorsman who has carefully observed the weather his entire life, and he's convinced the planet is headed into an extended cooling trend. He may be correct. After all, according to NOAA, summer temperatures were the 34th coolest on record since 1895, with Boston recording its sixth-coldest June since 1872, and New York City establishing its eighth-coldest June since 1869. Michigan, Wisconsin, Iowa, and Minnesota notched their coldest summers ever. A double dose of uncanny weather hit in October as it was our nation's wettest October on record, and the third coldest.

Checking the headlines at Drudge Report this week have only highlighted the fact this winter has been a doozie:

Winter Could Be Worst in 25 for USA
Britain's Big Snow Shuts Cities
Elderly Burn Books for Warmth
Cold Snap Spurs Power Rationing in China
Midwest Sees Near-Record Lows, Snow By The Foot
Miami Shivers from Coldest Weather in Decade
Weekend Freeze Looms for Gulf Coast

And to think that in 1989, the Director of the United Nations' Environmental Program declared, "Entire nations could be wiped off the face of the earth by rising sea levels if the global warming trend is not reversed by the year 2000."[314]

Well, congratulations, earthlings! The trend has reversed. Since 1998, there has been no warming of the earth's atmosphere, and there has been an undeniable global cooling of at least one-third of a degree Fahrenheit (.2 Celsius) since 2007.[315] The theory of anthropogenic global warming is not true. In fact, I've also just checked the latest carbon dioxide report from the observatory at NOAA's Earth System Research Laboratory at Mauna Loa, Hawaii, where they have consistently measured CO_2 for the last 50 years. In 1998, for every one-million atmospheric particles they collected, 366.87 were CO_2. Since then, the number of carbon dioxide particles has increased by a meager 18.67 particles per million, an addition so slight you could never notice it unless you were told. And, just as we witnessed during the cooling trend from 1940 to 1970, since 1998, as carbon dioxide levels slightly increased, temperatures have *declined*.

The global whiners need a new theory, and hopefully one that focuses on that big, uncontrollable ball of boiling plasma in the sky. In fact, the whiners don't want you to hear what I'm about to tell you.

---

314 Quote from Noel Brown, Director United Nations' Environmental Program (UNEP), "Greenhouse Warming, Nations May Vanish, U.N. Says," *Miami Herald,* July 5, 1989.
315 Based on "UAH Globally Averaged Satellite-Based Temperature of the Lower Atmosphere, 1979-June 2009."

LOOK TO THE SUN

The anthropogenic argument is as egocentric as Ptolemy's Law, which insisted the universe revolved around *man*. Try as we may, humans are incapable of altering the earth's massive atmosphere, simply because the forces at work are too mighty—especially the force of the sun.

To illustrate, let me present one last science lesson. The sun is a typical star, with a diameter of approximately 865,000 miles (nearly 10 times larger than that of Jupiter). Its core is burning at an unfathomable 29 million degrees Fahrenheit, while its surface sizzles at about 10,000 degrees. As the earth spins through its perfect 24-hour cycle and rotates around the sun annually, the tilt of the earth's axis, relative to its plane of travel around the sun, causes our four seasons (the hemisphere closest to the sun is in summer, while the opposite hemisphere is in winter). If the earth's axis were tilt-less, or straight up and down relative to the orbital plane, there would be no seasons at all, since every point on the earth would receive consistent amounts of sunlight each day of the year. Interestingly, the current tilt of 23.4 degrees is in a state of gradual change, shifting between 21.5 to 24.5 degrees every 41,000 years. This is known as the "obliquity cycle." These changes in pitch obviously impact the seasonal weather, with summers and winters becoming more extreme as the hemispheres lean closer or farther from the sun. Presently the earth is in a decreasing obliquity cycle, (moving back toward a lesser cant of 21.5 degrees), which will eventually produce cooler summers and milder winters. The obliquity cycle can't be stopped.

Next there is the "precession cycle." While the obliquity cycle impacts the *tilt* of the earth's axis, precession affects the *direction* of the tilt. It's a bit more challenging to imagine, but essentially, over the course of about 22,000 years, the axis of the earth subtly wobbles from side to side—like a top during its slowing seconds of spinning. This, too, has an impact on the atmosphere, forcing major global climate changes every 11,000 years. The current wobble has placed the southern hemisphere closer to the sun, allowing for slightly warmer summers, and cooler, snowier winters. The increased snow and ice in the winter have had quite an effect on Antarctica, lowering the average annual overall temperatures there. The precession cycle can't be stopped.

Finally, there is the earth's orbit around the sun, which is not a perfect circle, but rather, an ellipse. Over the course of some 100,000 years, the ellipse varies to such a degree that at one point the earth is a maximum of 94.5 million miles from the sun, whereas during its closest point of orbit, the earth and sun are 91 million miles apart. Obviously, at its farthest stage, there is slightly less solar radiation reaching the earth, resulting in a lower average temperature; at its closest proximity, the earth receives slightly more radiation, resulting in more habitable temperatures, as we are experiencing today. Once the earth passes from its current optimal position, the planet will be in a frigid phase that will stretch thousands of years, perhaps leading to a major ice age. This cycle, known as "eccentricity," is also unstoppable.

These three cycles were popularized by Milutin Milankovitch, a Serbian geophysicist, who died in 1958. Milankovitch is best known for his theory of ice ages in relation to the three cycles, now referred to as the Milankovitch cycles. Depending on the conjunction of these unstoppable cycles, one can clearly see that the earth's climate will be, and certainly has been in the past, radically altered *naturally*. Even if the sun's output of energy never varied, the amount of solar radiation reaching different areas of the earth would still change because of the uneven way in which the earth moves around the sun, allowing for both great ice ages and lengthy tropical outbursts.

However, the sun has an unstoppable cycle of its own. On the sun's surface are what we refer to as "sunspots." These are regions of increased magnetic energy, usually about the size of the earth. Within those sunspots are solar flares, or huge eruptions of boiling plasma. Solar flares emit X-rays and magnetic energy that bombard the earth with geomagnetic storms, often witnessed in the form of the eerie Northern (and Southern) Lights. During sunspot maximums, the earth can experience a disruption in electric power grids and radio and satellite transmissions. Though anthropogenic global warming advocates hate to admit it, variations in sunspot activity *are* associated with obvious increases and decreases in energy output from the sun, thus, potentially raising and lowering temperature accordingly. Solar cycles typically last about 11 years, but there have been unexplained episodes in which they've endured for decades. In fact, during the seventeenth century, the sun plunged into a 70-year period of spotless quiet known as the Maunder Minimum, a duration which coincided with the peak of the Little Ice Age.

The last solar cycle climaxed about the year 2000. Perhaps not coincidently, we have been experiencing a slight cooling since. Currently, the sun has entered significant ebb, with the least amount of sunspot activity since 1913. According to NASA, "There were no sunspots observed on 266 of the year's 366 days. To find a year with more blank suns, you have to go all the way back to 1913, which had 311 spotless days. Prompted by these numbers, some observers suggested that the solar cycle had hit bottom in 2008. Maybe not. Sunspot counts for 2009 have dropped even lower."[316]

"In our professional careers, we've never seen anything quite like it," says Dean Pesnell of the Goddard Space Flight Center. "This solar minimum has lasted far beyond what we predicted in 2007."[317] Almost unanimously, solar physicists link the lull in solar flares with the current cooling. A significant reversal will no doubt raise temperatures accordingly.

Between the incessant Milankovitch cycles and the sun's own unrelenting perturbations, we can clearly see that mankind's ability to alter the climate of the earth is pathetic by comparison. Manmade global warming *is* the biggest scam in history.

HOW, THEN, SHALL WE RESPOND?

Recently, I received a delightful, handwritten letter from a gentleman named Allen who wanted to tell me about his hero. Allen is 85 and his hero, Stan, 84. In 1945, while fighting the Nazis in the bloody Battle of the Bulge, Stan was struck with a bullet and paralyzed. Despite his fate, he went on to marry a wonderful nurse, adopt a beautiful daughter, get his MBA, and become a vice president for Bank of America. "Never once," Allen wrote, "did Stan ever complain about *anything*."

Allen served with Stan in the 11th Armored Division during that historic battle, describing himself to me, writing, "I never received a Purple Heart. I was a survivor, not a hero."

---

316 Dr. Tony Phillips, "Deep Solar Minimum," NASA press release, April, 1, 2009, http://science.nasa.gov/headlines/y2009/01apr_deepsolarminimum.htm.
317 Tony Phillips, "New Solar Cycle Prediction," *Sign of the Times*, May 27, 2009, http://www.sott.net/articles/show/185466-New-Solar-Cycle-Prediction-Fewer-Sunspots-But-Not-Necessarily-Less-Activity-who-knows-.

I phoned Allen and invited him to be a guest on my radio program. He accepted, and we had a most remarkable and enlightening conversation. Allen described in detail Stan's character, the courage of his fellow soldiers, the showmanship of General Patton, and the gut-wrenching liberation of the Mauthausen concentration camp.

Toward the end of our phone conversation, I asked Allen about the present condition of America as he saw us racing toward unprecedented government expansion, unpayable debt, and liberty-sapping policies.

"You've lived a very full life, Allen. You've seen it all. You've dodged Nazi bullets, shed tears at Mauthausen, returned home a hero, raised a family with your beautiful wife, and now you are witnessing your very own country being so deceived."

"Brian," Allen said without hesitation, "it's 1938 and the U.S. is Germany."

Allen's sobering statement later caused me to reflect on a concentration camp survivor I had read about by the name of Martin Niemoeller.

Niemoeller was born in Germany in 1892. During World War I he served as a successful submarine commander and after the war was a celebrated hero. Niemoeller followed the footsteps of his father and became a pastor and family man. Like many patriotic Germans of his day, Niemoeller initially supported the Nazi party and its charismatic leader, Adolf Hitler, especially because Hitler stressed the importance of Germany's Christian heritage and its role in the renewal of national morality and ethics. Hitler was a brilliant politician who engaged Germany with a whirlwind campaign in 1930 unlike anything ever seen before. Germany was in the grip of a global economic depression, and the German people suffered from poverty and uncertainty, amid increasing political instability. Hitler traveled the country delivering scores of upbeat speeches promising "freedom and bread," conducting town hall meetings, shaking hands, signing autographs, posing for pictures, and kissing babies. His public relations chief, Joseph Goebbels, orchestrated Hitler's every move, creatively organizing dramatic evening torchlight parades, plastering Hitler posters everywhere, and printing millions of copies of special editions of Nazi newspapers.

Hitler's campaign appearances were carefully staged events with audiences always kept waiting, deliberately allowing the tension to build to a crescendo of proud and blaring military music and solemn processions of brown-shirted youth carrying golden banners emblazoned with the Nazi insignia. Finally, the stage set, Adolf Hitler would appear amid victorious shouts of "Sieg Heil!" and near idol worship.

After three years of relentless political maneuvering, on January 30, 1933, Adolf Hitler was named Germany's chancellor. "I will employ my strength for the welfare of the German people, protect the Constitution and laws of the German people, conscientiously discharge the duties imposed on me, and conduct my affairs of office impartially and with justice to everyone," pledged the new leader.

His oath and his many promises quickly proved hollow. Later that same year, the first concentration camps were established, with political dissidents and then Jews sent to them. Niemoeller soon saw through the Nazi party's dangerous charade and began to speak against the government from his pulpit. The righteous indignation in his sermons resonated with many Germans and provoked the Nazi authorities to arrest him on July 1, 1937. He was sentenced to seven months in prison and fined. After his release, he refused to remain silent, so Hitler ordered him arrested again in 1938. He spent the next seven hellish years in concentration camps. Unlike many others in the world, Niemoeller was *not* deceived.

Hitler's rise to power was unparalleled, and surrounding countries soon were concerned about the dangers of Nazi expansion. After a highly publicized meeting between Hitler and Prime Minister Neville Chamberlain of England, Chamberlain heralded to the world, "I believe it is peace in our time." Soon after the prime minister's words were gobbled up and then regurgitated by the press, Hitler unleashed a ferocious campaign of terror upon the European continent and beyond.

Niemoeller later famously lamented that he didn't do enough to stop Hitler's reign of evil: "Then they came for the Jews, but I was not a Jew, so I did not speak out. And when they came for me, there was no one left to speak out for me."

If good people, like Niemoeller, wished they had vehemently spoken out as the most horrific lie in the history of the world was being carried out before their very eyes, do *we* have any excuse for not confronting those promoting this shameful scam of man-caused global warming?

No, we don't. Our country is in the throes of change, evil change designed to dismantle us and force us into a global political and economic system with unprecedented speed and on widespread levels not seen in this country since the days of America's Revolution. Perhaps we're past due for another. Personally, I refuse to remain silent about the literal betrayal and destruction of our still great nation, albeit a nation crippled by its own government.

If, indeed, environmental temperatures continue to decrease, or if global temperatures fail to rise, will the broad global warming policies being put into place today be repealed? Will we, the populace, again be allowed inexpensive and plentiful energy to adequately supply our needs and wants? Will the energy taxes be repealed? Will the government bureaucracy created to combat climate change be dismantled and restrictions lifted?

And what about other socialist issues such as universal health care, invasion of property rights, intrusions on the Second Amendment, freedom of speech on the radio and other media, and even forced toleration of practices once deemed immoral? Are we going to remain silent?

Every American citizen is going to have to make that decision for his or her self. It would do us well in making that decision, though, if we heeded the warning of founding father Thomas Jefferson: "All tyranny needs to gain a foothold is for people of good conscience to remain silent."

It is crucial to our nation's survival that we understand that our own government is pulling a *coup d'état* right under our very noses by recklessly disregarding the law of our land, the U.S. Constitution, and effectively usurping the power of "we," the American people. Tyrannical conditions are quickly becoming frighteningly similar to the conditions our nation experienced in its gestation period with King George's England in 1776.

I realize that the thought that this kind of thing *can* happen and *is* happening in the greatest nation in the history of the world is a tough pill to swallow, because, after all, these kinds of things are supposed to only happen in other, communist/socialist/fascist countries, *not* in the country of the land of the free and the home of the brave. Curtailing this stunning loss of freedom is going to take a brave and godly response from those who recognize the signs of the dark times in which we're living—a resolute response on par with that of our courageous forefathers, Martin Niemoeller, Allen, Stan, and all the great men and women of history who stood against tyranny.

May God give us the grace to do what is right.

INDEX